Fermentation *for Beginners*

Fermentation *for Beginners*

THE STEP-BY-STEP GUIDE TO FERMENTATION AND PROBIOTIC FOODS

DRAKES PRESS

For general information on our other products and services or to obtain technical support, please contact our Customer Care Department within the U.S. at (866) 744-2665, or outside the U.S. at (510) 253-0500.

Drakes Press publishes its books in a variety of electronic and print formats. Some content that appears in print may not be available in electronic books, and vice versa.

ISBN Print 978-1-62315-256-7 | eBook 978-1-62315-306-9

Contents

PART ONE

Introduction to Fermentation

Introduction

Humans have long loved, even craved, fermented foods and drinks. Without fermentation, we'd have no wine or cheese. We also wouldn't have salami, prosciutto, bread, beer, yogurt, or sourdough. Some fermented foods are very pungent, like kimchi or sauerkraut. Some have peculiar textures, ranging from slippery to sticky. We love these stinky, pungent foods. Perhaps we love them because in the past there was no other choice; sauerkraut and other fermented foods were important to keep us alive when winter made farming or hunting impossible. Perhaps we love them because our bodies crave them to stay healthy and strong.

The fermentation process has the effect of preserving foods, so it was an important way for humans to feed themselves when fresh foods were scarce. It also converts starches in foods to make them more easily digestible. Eating fermented foods improves our digestion by introducing living probiotic cultures into our guts and helps us absorb more of the nutrients from the foods we eat.

Fermented foods have been changed by microbial activity. The word "fermentation" indicates that this change is beneficial. Some bacteria are benign, some are beneficial, and some are harmful. When a microbial system is in balance, whether in our gut or in our food, harmful microbes are kept in check by beneficial ones. When the system is disrupted, harmful microbes can get the upper hand and flourish at the expense of beneficial ones. That is the difference between a fine wine and moldy grape juice, or between a delicious cheese and spoiled milk.

CHAPTER ONE

Good Reasons to Ferment

People who live in Asian countries like Korea and China, cultures where fermented foods like pickles and yogurt are common, are healthier and live longer, and the evidence that fermented foods are beneficial to our health is growing.

Fermented foods are not only good for us, but also delicious. Fermentation creates complex flavors—fizzy and tart, savory and satisfying—that no other kind of food has.

When you make your own fermented foods, you will derive the greatest nutritional benefit from them. Commercial pickles and other fermented products may not have been made through true fermentation; they may have been only been cured in vinegar. The flavor, although similar, is rarely as robust as an artisanal or homemade version. Even foods that are fermented using traditional methods can lose their potency, since the process of pasteurization can kill off most of the beneficial bacteria.

FERMENTING BOOSTS HEALTH

There are many health benefits when you eat fermented foods on a regular basis. These foods are an aid to digestion. They restore the colonies of healthy bacteria (flora) in the gut, another name for the alimentary canal, which is the place where food is digested and absorbed into the human body. This includes the stomach and the intestines. Studies have shown that fermented foods can lessen or reduce allergies. They reduce plaque bacteria in the mouth, resulting in fewer cavities and healthier gums. When people include fermented foods in their diets, they tend to lose weight more easily. And fermented foods may even help prevent such serious diseases as arthritis and atherosclerosis.

Why does balancing flora in the gut matter? When things get out of balance, the natural defense against harmful bacteria is not as robust. That is when we can start to experience a wide range of problems, from headaches to diarrhea to allergies. It is very easy for us to throw our balance out of whack simply by living an average, modern lifestyle. Poor food choices, emotional stress, poor sleep habits, even environmental conditions can easily upset the flora in the gut. Introducing fermented foods, with their beneficial living microorganisms, can bring things back into balance.

FERMENTING CREATES HEALTHY PROBIOTICS

Fermentation depends on thriving colonies of friendly microbes to do its work. Once the microbes find themselves in the friendly environment of a food with plenty of water and nutrients, they start to grow and reproduce. As they consume the sugars in the food, they produce alcohol and carbon dioxide. This changes the flavor and texture of the food. When we eat them, they go directly to the gut, where they encourage the growth of even more healthy bacteria.

The word "probiotic" comes from Greek words that mean "promoting life." A probiotic is a living organism. That is why you see the words "live" or "living" on the labels of your favorite yogurt. In fact, most fermented foods contain probiotics and prebiotics. Prebiotics do not actually contain the living organism themselves; they do create a favorable environment for healthy microbes. Both probiotics and prebiotics work together to promote digestive health and boost the immune system.

As their names indicate, probiotics and antibiotics perform opposite tasks. We take antibiotics in order to kill or weaken harmful bacteria that have increased and caused an illness. We eat probiotics because they promote the growth of beneficial bacteria that support the immune system.

FERMENTING SUPPORTS IMMUNITY

Our immune system has evolved to protect us from a wide range of dangers in our environment. The first line of defense is the digestive system, which not only takes in nutrients but also filters toxins out. It contains nearly 80 percent of the body's bacteria, nearly 100 trillion bacteria. If you could separate all the bacteria cells from the body's cells, the bacteria cells would outweigh the body cells.

The gut is the largest organ in the immune system and responsible for nearly half the body's immune response. There are the same number of neurotransmitters in your gut as in your brain, and this has led scientists and researchers who study the immune system to refer to it as a secondary nervous system. Over time, it has become apparent that the microbes in the gut communicate with immune cells and cause them to perform in a specific fashion. In addition to promoting good digestion, these microbes also support and may even activate the immune system.

Inflammation is associated with a wide range of diseases. Stiff and swollen joints that come from arthritis are one example. Cardiac disease is another. It appears that the probiotics in fermented foods can reduce this inflammation by communicating directly with the cells that cause the inflammation in the first place.

FERMENTING AIDS DIGESTION

A healthy digestive tract can do its job easily. As it breaks down the foods we eat, it absorbs the nutrients and water that our bodies need. It also filters out toxins and other harmful substances in the food.

Sometimes, the foods we eat can interfere with our bodies' ability to digest them. You've probably heard some foods described as "easy to digest." What that means is that the body does not have to work as hard to extract the nutrients and filter out the toxins.

The process of fermentation removes or inactivates substances in foods that might otherwise give us digestive trouble. One such substance is known as phytates, which have been referred to as antinutrients. Instead of providing nourishment, phytates prevent us from digesting foods.

That is why phytate-rich foods like grains, nuts, seeds, and legumes need to be processed in some way before we can get any nourishment from them. Although we often think of processing in negative terms, the type of processing that we are referring to here includes primarily drying, grinding, and cooking. Fermentation plays a hand even here; that is why grains are allowed to "age" after they are ground.

Fermented foods are, in effect, already digested by the time we eat them. The bacteria in cultured dairy foods have already consumed the lactose in milk, which explains why some lactose-intolerant individuals can consume yogurt or cheese without a problem.

FERMENTING PRESERVES FOODS

All of us have probably left something in the refrigerator too long or out on the counter overnight. The food has definitely changed. It might be moldy, or withered, or even frothy. You might see molds of various colors or the food may look as if it has simply melted. And then there is the smell. A food that has rotted smells truly awful. A food that is fermented usually smells strong, but it still smells appetizing.

Fermenting strikes the balance between creating an environment that allows the good microbes to thrive and that slows down or stops the bad microbes from turning food into a science project. If you control the process successfully, you end up with something edible.

Many fermented foods are often combined with a significant amount of salt. Salt destroys the microbes that might cause foods to spoil or putrefy. With those microbes out of the way, there is less competition for food, so the healthy microbes can take hold.

FERMENTING IS TRADITIONAL

It isn't hard to understand why our forbearers first became interested in harnessing the process of fermentation. It was a question of survival. If there is one thing our current love affair with seasonal eating has done successfully, it is to remind us of what our ancestors always knew: If you are going to make it from one season to the next, you need to prepare. You need a place to shelter you, clothes to keep you warm, and food to keep you strong, even when the wild game is gone, the heifer has gone dry, and the fruit trees are finished producing. This is our human patrimony, passed down to us by those who learned to harness microbes and make milk into cheese, grapes into wine, and anchovies into fish sauce.

Every cuisine on earth has a tradition of fermented foods. Egyptians controlled yeast to make both bread and beer. Fruit juices left to ferment have been made into wines, cordials, and tonics for thousands of years; they are a part of the culinary tradition anywhere there are vineyards or orchards. The Romans were famous for their garum, which was essentially fermented fish guts. Norwegians realized that the salmon they buried and left behind as they continue to fish upstream was different, but delicious. Adding salt to the salmon before they buried it pulled the water out of the fish that would otherwise make it rot. Today, we have gravlax with cream cheese on a bagel. The barrel of vegetable scraps left to age through a cold Russian winter was the original borscht, nothing more than fermented ends and cores of beets and cabbages and onions. Today, fermented beet juice, known as kvass, is acknowledged as a rich, healthy drink. Long ago in Korea, burying cabbage to last through the winter ultimately resulted in what we know as kimchi. In Europe, sauerkraut was the Western equivalent.

Traditionally prepared pickles were always preserved in salt; the lactic acid produced during fermentation gave them their tartness. Commercially prepared pickles today are rarely made with salt; instead, the vegetables are preserved in vinegar. If you are looking for traditional pickles in the market, read the label. If you see vinegar as a main ingredient, they are not fermented pickles.

FERMENTING LETS YOU EXPERIMENT

Foods that are fermented using traditional techniques have a fantastic flavor. They are quite easy to make, although you do have to allow time for the fermentation to happen. That said, fermentation is also a craft. Every batch you make will have its own character because every environment is different. You don't have the same conditions every day, even in your own home.

Supermarkets have started stocking bottles of drinks like kombucha and kefir. With very little effort or expense, you can make these drinks and a host of other fermented foods. The process takes several hours or several days, but your only task is to maintain a friendly environment and perhaps feed the mixture from time to time.

Once you know the basics, you can ferment virtually any vegetable or fruit, make sodas from teas or juices, and even create your own fermented beans for miso and tempeh.

CHAPTER 2

An Overview of Fermentation

Our very survival as humans is linked to fermentation. Evidence of controlled fermentation goes back to the earliest agricultural settlements in Europe, Asia, and South America. Like all basic human skills, fermentation was developed through trial and error. The methods earlier generations developed were based on the evidence before them: what temperature was most favorable, what amount of salt or sugar was needed, what resulted when foods were buried or submerged, and how long it takes for fermentation to stop.

Today, we have a better understanding of what happens during fermentation and how the food is being changed. It is helpful to think of fermentation as an ecosystem, with a variety of different organisms all doing their best to survive. The types of species in the system have a direct effect on the chemical composition of the environment.

Fermentation depends on the dynamics of the population (the types of microbes present in the system), the limits to the source of nutrition, and competition for those nutrients. The natural environment of the fermentation system determines what the fermented food will taste, look, sound, and feel like.

THE FERMENTATION PROCESS

The organisms responsible for fermentation are in the atmosphere, whether we specifically add them or not. Uncontrolled fermentation, when wild yeasts and various strains of bacteria are introduced to foods, can result in something delicious but often results in something spoiled and inedible. Learning to control the process has had numerous benefits for humans. It meant that human beings had enough to eat and drink over the course of a year (don't forget, there was no refrigeration). Control of the fermentation ecosystem was already being practiced by human cultures long before written records were kept. Our ancestors were adapted to their environment and its resources and gave us a rich selection of foods that have been preserved following traditions that are as old as humankind.

As we have refined our techniques and tools over time, the process has become easier to control as well as easier to understand. There are a variety of tools used in the fermentation process, ranging from using one's senses to judge color, flavor, or texture to more accurate tools and technologies, such as thermometers, hydrometers, metering tubes, and siphons.

Microbes are everywhere in the environment, including large systems like oceans or the soil and small systems like our guts. When a given type of microbe has plenty of the things it needs, like food and water, and relatively little of the things that are harmful to it, like very high or low temperatures or acid levels, and not too much competition, the microbe thrives. As it grows and reproduces, it converts sugar into alcohol and acids.

There are two types of organisms that play a key role in fermentation: fungi and bacteria. In general, bacteria produce acids, while fungi produce alcohol. Some foods are fermented with a combination of fungi and bacteria.

Fungi are yeasts and molds that are used to produce wine, beer, cheese, and bread. They produce ethanol, a type of alcohol, and give off carbon dioxide (producing the air pockets in bread and the "eyes" in some cheeses). Even though the alcohol is not the point of fermenting a dough, it does add specific flavors to the bread.

The strain of bacteria that is most often associated with fermented foods is *Lactobacillus* (used in pickles, cheeses, and fermented sausages); this strain gives fermented foods a specific tanginess that is the result of the lactic acid it produces. A different strain of bacteria, *Acetobacter*, produces acetic acid from the alcohol in wine and hard cider, giving us vinegar.

The requirements for controlled fermentation are a desirable strain of microbes, the right type of organic material, and the ability to create the optimum environment for fermenting.

MANAGING THE FERMENTING ENVIRONMENT

Some microbes cause foods to rot and putrefy. Others compete with the desirable microbes, preventing fermentation from taking place. The goal of food fermentation is to encourage the good microbes that turn cucumbers into tart crisp pickles or grape juice into wine while discouraging or destroying the bad microbes that create funky molds and foul odors. By repeatedly allowing the same sequence of bacteria to grow in foods, bakers, brewers, cheesemakers, vintners, and food preservationists are able to produce foods and beverages with consistent flavors and textures.

INTRODUCING OR ENCOURAGING A DESIRABLE STRAIN

Some types of fermentation occur because wild yeasts and bacteria are present on the surface of food or in the environment. This type of wild fermentation is still used today for pickling vegetables or making bread.

In other instances, a specific type of starter is called for. They might be cultures or starters, or they may be purified forms that can be used to create a very specific product.

The goal is to encourage the good microbes that give us lactic acid for tart pickles while discouraging or destroying the bad microbes that might result in funky molds.

Different organisms have different requirements. Some need plenty of oxygen all the time (aerobic), some cannot be exposed to oxygen (anaerobic), and some get along either way (facultative). Likewise, most have a preferred temperature range: very cold (cryophilic), mid-range (mesophilic), or very warm (thermophilic). Knowing an organism's favorite temperature and being able to maintain it throughout fermentation is one of the most important keys to success. Some organisms like the environment to be fairly acidic, some prefer an environment that is alkaline, and some prefer a neutral environment that is neither acidic nor alkaline.

Salt is a key player not only because of the way it changes the amount of water that is available in the environment. Adding salt also changes the acid level (pH) of the ecosystem. Some bacteria, like *Lactobacillus,* thrive in the conditions a salty environment creates. Once they are established, they can also change the pH of the environment.

Lacto-fermentation, the procedure used to produce pickled vegetables, fruits, meats, and dairy foods, is an anaerobic process. By depriving the bacteria that cause foods to rot of oxygen, we create an environment that favors other bacteria. The lactic acid that *Lactobacillus* bacteria produce gives pickles their characteristic taste and turns milk into yogurt or cheese.

STOPPING OR SLOWING FERMENTATION

At some point in the fermentation process, equilibrium will be reached. The feverish activity of the first stages of fermentation will stop or slow significantly. Some ferments change from cloudy to clear as the dead cells settle to the bottom of the fermentation vat.

Modern refrigerators are generally slightly colder than traditional cold storage or cellar temperatures (around 37°F, just a few degrees from freezing). While fermented foods stored in the refrigerator continue to ferment, it happens much more slowly. Cold storage temperatures maintained in caves or cellars (40° to 50°F) also slow fermentation; in addition, these temperatures are appropriate for some foods or beverages like hard cheeses or wine that need additional time and aging in order to develop the desired flavor, texture, color, and aroma.

Another way to stop or slow fermentation is to raise the temperature of the food to about 180°F. This heat treatment pasteurizes the food, killing off the microbes responsible for fermentation. That means that you no longer get the benefits of the living probiotics, of course, but you will have a more shelf-stable item.

FERMENTING AND SPOILING

Saftey is a major concern for lacto-fermented foods. But fermented foods are far more likely to resist spoilage than other foods because their environment is unfriendly to those types of microbes. Fermented foods can spoil, of course, but when they do, your nose and eyes are a sufficient early warning system. By contrast, even cooked foods that look, smell, and taste just fine can still harbor the pathogens that produce food-borne illnesses.

The process of fermentation is complex and involves many variables. Food scientists are still investigating its myriad aspects. Regardless of how many interlocking systems there may be at work at a cellular level, the cook's work when it comes to fermenting foods is direct and straightforward. The waiting is probably the hardest part.

CHAPTER 3

How Fermentation Works

You can ferment almost any food as long as you have the right combination of microbes and a rich source of nourishment. The basic ingredients for fermentation are rarely exotic or difficult to find. You probably have most of them in your kitchen right now.

BACTERIA

Lacto-fermentation requires the presence of *Lactobacillus* bacteria. There are many different strains of *Lactobacillus*. Some prefer a certain type of food; some work best with vegetables, some with dairy, some with fruit, and some with alcohol. *Lactobacillus* bacteria are anaerobic, and they convert sugar to alcohol in a process known as glycolysis. The bacteria also give off gas (carbon dioxide) as part of the conversion process, which is why containers for fermented pickles or kraut need to be opened periodically to release gas or call for the use of an airlock to vent the gas as it builds up but prevents oxygen from entering. *Acetobacter* bacteria need oxygen in order to turn alcohol into vinegar.

As the bacteria grow in the nutrient-rich environment, they divide—one cell becomes two, two cells become four. When cell division is occurring rapidly, the overall population of the bacteria grows at a furious pace. Therefore, it is easy to understand why you need only a small amount of bacteria to ferment a large amount of food.

Some of the bacteria for lacto-fermentation are fairly widespread in the environment. You simply let the bacteria that are already present multiply by giving them the right environment. Or you introduce other types of bacteria, such as adding buttermilk or yogurt to whole milk to make yogurt.

Some bacteria can thrive through many generations as long as they are fed regularly. This is especially true of cultures that start with wild ferments. Powdered and freeze-dried forms of bacteria for fermentation are rarely effective after more than two or three ferments.

FUNGI

Yeast and molds are types of fungus that can ferment foods. The process is slightly different, but the end result is the same. While bacteria reproduce by simply splitting apart, yeast reproduce through a process called "budding." A small bud emerges on the parent cell. The parent cell nucleus splits in two. One nucleus moves into the bud cell, which continues to grow until it is large enough to split from the parent cell. When temperatures, moisture, and nutrients are available, yeast can reproduce as dramatically as bacteria.

Molds are used less frequently to ferment foods, although they do play a role in the production of cheeses, certain wines, and specialty foods like the so-called corn smut, or huitalacoche, of Mexico. The mold used to make Roquefort is a type of penicillin, while the *Botrytis cinerea* mold that attacks grapes on the vine is known as "noble rot" because it produces wines of exceptional structure and bouquet.

Yeast and molds are naturally present everywhere. Because they are so prolific, the native fungi will crowd out other strains. That is why a sourdough starter that moves from San Francisco to New York eventually loses its typical San Francisco tang in favor of the local population. It is also why some wine or beer makers prefer to use strains of yeast developed especially for wines and beers.

MOTHERS, SCOBYS, KEFIR GRAINS, AND WHEY

A mother will develop in any raw or unfiltered vinegar eventually. The bacteria that is responsible for turning wine into vinegar gets into the wine via a ride on the vinegar wasp and once in, they get to work eating the alcohol and turning it to vinegar. Eventually, a colony of cellulose and bacteria starts to form into what is best described as a blob. This blob is called the "mother." The appearance of a mother can be disturbing if you aren't expecting to find one inside your vinegar. It looks grayish and slimy and is covered with spots and blotches.

A mother will continue to live almost indefinitely, as long as it is fed periodically with fresh wine. Eventually, a second blob, or a baby, will form on top of the mother. The older mother generally begins to sink as it becomes less active. Older mothers are less potent, but they are almost always potent enough to be given to someone who wants to start brewing their own vinegar.

Some vinegar-making enthusiasts suggest using a specific type of mother for each wine (one for red wine, one for white, and one for cider). Others don't think it makes much difference.

SCOBY stands for "symbiotic culture of bacteria and yeast." A SCOBY is the gelatinous blob that forms on the surface of sweetened tea as it ferments. The specific bacteria and yeast feed on the sugary tea, creating kombucha. At the end of fermentation, the kombucha tastes tart and refreshing, quite unlike the sweet tea that it began as. A new SCOBY grows with each batch of kombucha you brew.

Kefir grains are similar to a SCOBY. They also contain both yeast and bacteria. Some are meant for dairy, while others can be used for nondairy liquids (including nondairy milks). When they are used to brew a batch of kefir, the grains grow. You might eventually find yourself with more than you can handle. That's when you have to decide whether to share or compost.

Whey is the liquid part of milk that drains away from curds. This liquid contains the *Lactobacillus* bacteria used to ferment a wide variety of foods, including fruits, vegetables, and meats. Whey is not necessary for fermentation, but it can have the effect of speeding it up. Many fermenting enthusiasts opt for using a little salt instead of whey. Others feel that whey contributes to the pickle's nutritional value as well as its flavor. Our recipes offer both options. You can make your own whey by draining homemade or commercial yogurt or kefir as described on page 53. (If you are using commercial yogurt, use plain yogurt that contains live or active cultures.)

STARTERS

Traditional bread bakers around the world have always relied on prefermented doughs to help leaven breads. Each preferment has its own characteristics. Some are left to ferment for only a few hours and are very wet. Prefermenting the dough gives it enough time to develop rich flavors because the yeast is able to develop slowly.

Sourdough starters are similar to preferments in some ways. The main difference is that you use up the entire preferment when you make the final dough. When you are working with a sourdough starter, you use only a portion of the starter for your bread; you keep a portion of it alive by feeding it.

Specific flours are used to make sourdoughs for a wide array of global breads, from San Francisco sourdough boules and baguettes to a French miche or German rye, from Russia's buckwheat blini to the millet-based injera of Ethiopia.

Establishing and maintaining a starter is a commitment of time and energy, but there is no substitute for rich, appealing flavor it produces in bread.

SELECTING INGREDIENTS FOR FRESHNESS

Buying or growing foods that are additive-free and raised without chemical fertilizers and pesticides makes a lot of sense when you are preserving foods, especially if you will be including the skins. But there is no specific requirement that the foods you use must be organic. Fermentation has always been employed to preserve the foods that are in season. They can be in any state from immature to overripe, as long as they are not moldy or at the point of disintegration.

FRESH PRODUCE

When a food is at its peak and bountiful, that's the time to preserve it. Some foods have specific and very short seasons, like asparagus, garlic scapes, or strawberries. Others have longer seasons as different varieties of the same fruit or vegetable come to maturity, such as apples or tomatoes.

As a general rule, fruits and vegetables should be rinsed and trimmed to remove stems or bruises. One notable exception is grapes for making wine. Rinsing them would remove the yeasts that are present on the grape's surface.

SALT

The type of salt you use can make a difference in the way your ferments taste. Kosher salt is good because it has no additives. However, the grains are large and they often take more time to dissolve than salt with finer grains. Table salt, in addition to containing dextrose (sugar), contains anticaking ingredients and sometimes iodine, both of which can affect the flavor and appearance of your pickles. Canning and pickling salt is additive-free and very fine, so that it dissolves more easily. Sea salts vary greatly in terms of flavor and mineral content. In the end, the type of salt you choose is a matter of preference and price.

WATER

If your drinking water is treated with chlorine or fluoride, you should consider using filtered or bottled spring water for fermenting.

RAW MILK AND CIDER

In most parts of the country, selling raw milk and cider is illegal, unless the farm itself is certified to sell milk. Customers must go to the farm and purchase the milk directly from the farmer. There are some fermented foods that are difficult, if not impossible, to make without raw milk or cider. Clabbered cream, for instance, is nothing more that raw milk left to separate at room temperature. Pasteurized cider is missing the bacteria necessary to turn the fruit juice into alcohol. In general, however, any milk or cream that is not ultra-pasteurized or treated with preservatives can be used to make most cultured dairy foods. In fact, raw milk has some disadvantages if you plan to make it into yogurt. The bacteria in the raw milk can eventually crowd out the bacteria that turn it into yogurt. In that case, you must maintain a separate strain of starter made with pasteurized milk.

CHAPTER 4

Your Home Fermentation Laboratory

Many fermented foods can be made with tools you already own or that you can buy easily and inexpensively at the grocery or hardware store. Check local listings to see if there are any brewing or wine-making supply stores in the area. See Resources (page 127) to track down specialty items online.

Here are things you probably already have in your kitchen:

- Measuring cups
- Small digital scale
- Spoons
- Sieves and colanders
- Bowls
- Knives
- Graters
- Cutting board
- Vegetable peeler

If you don't have these on hand, you can find them at the grocery store:

- Cheesecloth
- Funnel

In addition, you'll need:

- Containers
- Lids

The following are helpful, but not always necessary:

- Thermometer
- Hydrometer
- Airlocks

- Weights
- Siphons
- Bottle brushes

Food processors, blenders, and other appliances can be helpful, but they are not necessary. You can almost always accomplish the same thing manually.

CONTAINERS FOR FERMENTING

The size, material, and shape of your container depend on what the food is and what quantity you are making. Glass and food-grade plastic are fine. Glazed ceramic is the traditional choice for pickling crocks.

Glass canning jars: These jars range in size from 4 ounces to 1 gallon. Wide-mouth jars are indicated when a recipe calls for them. Canning jars may have two-piece screw tops or a clamp seal with a rubber gasket.

Buckets and carboys: You need a large container for batches that won't fit easily in a canning jar or that may froth and form during the early stages of fermentation. Food-grade plastic buckets are one option. Carboys, large glass jars with narrow necks, are another. These containers come in a wide range of sizes, from 1 to 10 gallons (or even larger.) Plastic fermenting buckets should have airtight lids. Carboys should have plugs to seal the necks.

Pickle or kraut crocks: Ceramic crocks, glazed on the inside and without any scratches, are the traditional choice for making pickles. Some crocks have lids. You may be able to find crocks in antique and thrift shops. Check to be sure that there are no scratches in the glaze because scratches can harbor harmful bacteria.

WEIGHTS

When you put foods into a liquid brine, they will float. If they are above the brine, they are coming in contact with oxygen. To prevent that, add a weight to the top of the food. You can buy weights of various weights and in shapes meant to fit specific crocks or other containers. You can also make your own weight by putting a dish or plate on top of the food and setting a bottle filled with water on top of the plate.

AIRLOCKS

An airlock is a device that releases the gases that build up in the fermenting container so it doesn't explode. Monitoring the airlock gives you an idea of how fermentation is progressing. Early in the process, you see lots of bubbles. As fermentation slows down, the rate drops to less than one bubble per minute.

BOTTLES AND CAPS FOR FERMENTED BEVERAGES

You can use glass bottles for wine or beer that you already have, provided they are not chipped or cracked. Grolsch beer bottles with their clamp seals are especially popular. There is a variety of stoppers for fermented beverages, including corks, clamp-style stoppers, and caps. You can find these stoppers, as well as the tools that insert corks or attach bottle caps, at brewing supply stores (you may be able to use their equipment to cork or cap your bottles) or online.

THERMOMETERS

Dairy ferments are hard to do correctly without accurate thermometers. You should have an accurate instant-read thermometer (sometimes called a stem thermometer). A floating thermometer helps you keep track of temperatures during the fermentation process. They are made both with digital or analog displays.

HYDROMETERS

A hydrometer measures the strength of a liquid. When a brewer takes a reading and compares it to a chart, he or she can determine the amount of alcohol in the brew. To use a simple hydrometer, lower the bulb into the liquid and spin it to remove any air bubbles that would make it float higher. A gauge on the stem provides the reading, which is taken at the level of the liquid on the gauge.

SIPHONS

Getting a liquid out of the fermenting container without disturbing the sediment is easiest with some food-grade plastic tubing. This tubing can usually be found at hardware stores and appliance shops. A piece that is 6 to 7 feet long is usually adequate. Clamps hold the tubing closed, and bottle fillers act like a stopper to hold the siphon closed when you are filling bottles.

SANITATION

Cleanliness matters. Always wash all your tools and containers thoroughly in hot, soapy water. Use a bottle brush to reach inside containers and make sure that you clean out any crevices or corners completely. Rinse everything in hot water and let it air-dry.

For most fermenting projects, having clean hands, clean cutting boards, and a clean stem on the thermometer is sufficient.

Most brewers take the added step of sterilizing their equipment and containers. You can do this in several ways. Glass and metal tools can be sterilized in a bath of boiling water or with steam, but this may not be practical for large pieces or those made of plastic. In that case, you can make a solution of chlorine bleach and water (about 1 tablespoon of bleach per 1 gallon of water). Submerge the pieces in the solution and let them soak for 20 minutes. Some home brewers do this in a very clean bathtub. Rinse the pieces in hot water.

There are other sanitizers available that do not have to be rinsed off, sold under a variety of names. Campden tablets are one common product. Be sure to follow the instructions for the sanitizer you are using.

BASIC TECHNIQUES

Unlike cooking, which calls for mastering a wide range of techniques, fermenting primarily consists of combining ingredients and then letting time do the rest of the work. However, it is necessary to adjust the environment to achieve certain effects.

CREATING AN INCUBATOR

Foods that must ferment in a warm environment that ranges from 90 to 125°F call for an incubator. You will need to experiment at first to find the right spot in your home for your various ferments. You can use one of the following approaches:

- Set your ferment on a heating pad or a seedling pad.
- Put your ferment in a picnic cooler; add jars of hot water to keep the environment warm.
- Put your ferment on top of the refrigerator.

- Put your ferment in the oven with the pilot light on, if you have one, or your oven light on. If necessary, you can add pans of hot water to heat the interior of the oven.
- Put your ferment on top of an "always on" device like a cable box or a DVD player.

FERMENTING AT ROOM TEMPERATURE

There is really nothing special involved in fermenting at room temperature unless your rooms are cooler than 60°F or warmer than 80°F. Here are a few basic considerations:

Most ferments should be out of direct sunlight. Otherwise, the countertop is fine. So are the insides of cupboards.

Some ferments attract fruit flies. If this happens, put a dish or bowl nearby filled with fruit juice. Add a few drops of dish detergent and put it near the fruit flies. The juice will attract them, but when they land on the liquid, the detergent will make them stick.

Many ferments should not be moved or jiggled.

If you have different ferments going on at room temperature at the same time, keep them separated by about 10 feet.

PART TWO

Recipes

CHAPTER 5

Vegetables

Pickled vegetables are an important part of the meal in many cultures, from Japan to the Pennsylvania Dutch. A pickle spear accompanying a sandwich is such a familiar sight that most of us don't even wonder how it got there.

The recipes in this chapter are a sample of the wide variety of vegetables you can ferment. In addition, you can pickle cauliflower, squashes, pumpkin, cardoons, fennel, celery, artichokes, or tomatillos. In fact, apart from leafy greens and herbs, it's hard to think of a vegetable that you can't pickle.

When you are fermenting vegetables, make sure they are completely submerged in the brine. This may require the use of weights, as described on page 20.

Another tool you might want to use when preparing fermented pickles is an airlock. This gives you the ability to control the amount of oxygen that gets to the vegetables as they ferment and gives the gases the microbes are producing a way to escape before the pressure builds up too much. Information on airlocks can be found on page 21; to purchase airlocks, see Resources, page 127.

The recipes have been prepared in relatively small batches, so the fermentation takes place right in the canning jar the food will be stored in. Usually, the curve at the top of the jar is enough to keep the vegetables under the brine, as long as you've added enough liquid to cover them by about 1 inch.

If you are making large batches, you might prefer to let them ferment in a large crock. If you have weights and an airlock, these can be helpful too (although all is not lost without them). Once the initial fermentation is finished and you are ready to transfer the pickles to cold storage, you need to drain the pickle, but you also need to keep the brining liquid to add to the jars. Here's how to accomplish that task:

1. Set a colander inside a larger bowl. If you can suspend it over the bowl, so much the better.
2. Pour the pickle and the brine from the fermenting container or crock into the colander.
3. Let the pickle drain for several minutes.

4. Transfer the pickle to clean canning jars, packing the pickle into the jar to within 1½ inches of the top.
5. Pour enough of the reserved brine into each jar to completely cover the pickle. The top of the liquid should be about 1 inch from the top of the jar.
6. Cover the jars tightly with screw-top lids.
7. Transfer to the refrigerator or cold storage.

SOME SPECIAL NOTES ABOUT SAUERKRAUT AND KIMCHI

The recipes allow 3 days at room temperature for these fermented cabbage dishes. You might have heard that traditional recipes call for allowing them to ferment buried in the ground through the fall and winter before they are ready to eat. You can certainly eat our versions right after a fermentation period of 2 or 3 days, but the flavor continues to develop with longer periods of fermentation in cold storage. Eventually, the sauerkraut or kimchi may become too sour to eat.

Sauerkraut

MAKES TWO 1-QUART JARS

- WHEY STARTER (OPTIONAL)
- LACTO-FERMENT
- FERMENT AT ROOM TEMPERATURE
- FERMENTATION TIME: 3 DAYS

Sauerkraut is a traditional accompaniment to a wide variety of roasted and braised meats, especially sausages, beef, and pork. A famous Alsatian dish known as choucroute garnie is an elaborate presentation of the region's finest sausages braised on a bed of pungent sauerkraut mixed with apples and onions. Like all living foods, once you heat the sauerkraut to about 120°F, there will be no more living organisms in the dish. If you make your own sauerkraut, you might want to drink some of the juice. The liquid contains many important probiotics.

1 MEDIUM GREEN CABBAGE, CORED AND SHREDDED
1 TABLESPOON CARAWAY SEEDS
¼ CUP WHEY OR ADDITIONAL 1 TEASPOON SEA SALT
1 TABLESPOON SEA SALT

1. Mix the cabbage, caraway seeds, whey, and salt in a large bowl.
2. Pound the cabbage with a meat pounder or mallet until it releases some of its juices, about 10 minutes.
3. Pack the cabbage in two 1-quart wide-mouth canning jars.
4. Press down firmly until the juices rise over the top of the cabbage. The top of the liquid should be about 1 inch below the top of the jars.
5. Cover the jars tightly and let the sauerkraut ferment at room temperature in an area away from direct sunlight for about 3 days.
6. Transfer to cold storage.

Curtido

MAKES TWO 1-QUART JARS

- WHEY STARTER (OPTIONAL)
- LACTO-FERMENT
- FERMENT AT ROOM TEMPERATURE
- FERMENTATION TIME: 3 DAYS

This zesty Caribbean-style sauerkraut is a natural with grilled, roasted, or braised meats, especially pork and game. It makes a perfect addition to fish tacos, too.

1 MEDIUM CABBAGE, CORED AND SHREDDED
1 CUP SHREDDED CARROTS
2 MEDIUM ONIONS, QUARTERED AND SLICED VERY THIN
4 CUPS PINEAPPLE VINEGAR (PAGE 97)
1 TABLESPOON DRIED OREGANO
½ TEASPOON RED PEPPER FLAKES

1. Mix the cabbage, carrots, onions, pineapple vinegar, oregano, and red pepper flakes in a large bowl.
2. Pound the vegetables with a meat pounder or mallet until they release some of their juices, about 10 minutes.
3. Pack the vegetables in two 1-quart wide-mouth canning jars.
4. Press down firmly until the juices rise over the top of the vegetables. The top of the liquid should be about 1 inch below the top of the jars.
5. Cover the jars tightly and let the curtido ferment at room temperature in an area away from direct sunlight for about 3 days.
6. Transfer to cold storage.

Kimchi

MAKES TWO 1-QUART JARS

- WHEY STARTER (OPTIONAL)
- LACTO-FERMENT
- FERMENT AT ROOM TEMPERATURE
- FERMENTATION TIME: 3 DAYS

You can make your kimchi hotter if you like. Simply add a few more chilies or an extra ½ teaspoon or more of red pepper flakes.

1 HEAD NAPA CABBAGE, CORED AND SHREDDED
1 BUNCH GREEN ONIONS, CHOPPED
1 CUP SHREDDED CARROTS
2 HOT CHILIES, SLICED THIN
½ CUP GRATED DAIKON
1 TABLESPOON GRATED GINGER
3 GARLIC CLOVES, PEELED AND MINCED
½ TEASPOON RED PEPPER FLAKES
¼ CUP WHEY OR ADDITIONAL 1 TEASPOON SEA SALT
1 TABLESPOON SEA SALT

1. Mix the cabbage, green onions, carrots, chilies, daikon, ginger, garlic, red pepper flakes, whey, and salt in a large bowl.
2. Pound the vegetables with a meat pounder or mallet until they release some of their juices, about 10 minutes.
3. Pack the vegetables in two 1-quart wide-mouth canning jars. Press down firmly until the juices rise over the top of the vegetables. The top of the liquid should be about 1 inch below the top of the jars.
4. Cover the jars tightly and let the kimchi ferment at room temperature in an area away from direct sunlight for about 3 days.
5. Store the kimchi in covered containers in the refrigerator for up to 1 year. The flavor will continue to develop as the kimchi ages.

Garlic Carrots

MAKES TWO 1-QUART JARS

- WHEY STARTER (OPTIONAL)
- LACTO-FERMENT
- FERMENT AT ROOM TEMPERATURE
- FERMENTATION TIME: 3 DAYS

These spicy, garlicky carrots are the perfect topping for grilled fish or burgers. You can add them to coleslaw and other salads for a kick of heat.

4 CUPS SHREDDED CARROTS, TIGHTLY PACKED
2 TABLESPOONS MINCED GARLIC
¼ CUP WHEY OR ADDITIONAL 1 TEASPOON SEA SALT
1 TABLESPOON SEA SALT

1. Mix the carrots, garlic, whey (if using), and salt in a large bowl.
2. Pound the carrots with a meat pounder or mallet until they release some of their juices, about 10 minutes.
3. Pack the carrots in two 1-quart wide-mouth canning jars.
4. Press down firmly until the juices rise over the top of the carrots. Add more filtered water if necessary to completely cover the carrots. The top of the liquid should be about 1 inch below the top of the jars.
5. Cover the jars tightly and let the garlic carrots ferment at room temperature in an area away from direct sunlight for about 3 days.
6. Store the garlic carrots in covered containers in the refrigerator for up to 1 year.

Pickled Onions

MAKES TWO 1-PINT JARS

- WHEY STARTER (OPTIONAL)
- LACTO-FERMENT
- FERMENT AT ROOM TEMPERATURE
- FERMENTATION TIME: 3 DAYS

Handcrafted cocktails are all the rage. These delicious pickled onions give you bragging rights next time you make your world-famous Gibsons.

2 POUNDS PEARL ONIONS, PEELED
1 CUP FILTERED WATER, PLUS MORE AS NEEDED
¼ CUP WHEY OR ADDITIONAL 1 TEASPOON SEA SALT
1 TABLESPOON SEA SALT
1 TABLESPOON JUNIPER BERRIES
2 TEASPOONS WHOLE CLOVES
1 TEASPOON GREEN PEPPERCORNS
4 SPRIGS FRESH TARRAGON
1 CINNAMON STICK
1 SMALL WHOLE NUTMEG, CRACKED

1. Place the pearl onions in two 1-pint wide-mouth canning jars.
2. To make the brine, combine the water, whey, salt, juniper berries, cloves, green peppercorns, tarragon, cinnamon stick, and nutmeg in a bowl. Stir until the salt is dissolved.
3. Pour the brine over the onions, dividing it evenly between the jars. Add more filtered water if necessary to completely cover the onions. The liquid should reach to about 1 inch below the top of the jars.
4. Cover the jars tightly and let the onions ferment at room temperature in an area away from direct sunlight for about 3 days.
5. Store the pickled onions in covered containers in the refrigerator for up to 1 year.

Pickled Beets

MAKES TWO 1-QUART JARS

- WHEY STARTER (OPTIONAL)
- LACTO-FERMENT
- PREPARATION TIME FOR BEETS: ABOUT 3 HOURS
- FERMENT AT ROOM TEMPERATURE
- FERMENTATION TIME: 3 DAYS

Beets can stain your hands, so you may want to wear gloves when working with them.

12 MEDIUM BEETS
1 CUP FILTERED WATER, PLUS MORE AS NEEDED
¼ CUP WHEY OR ADDITIONAL 1 TEASPOON SEA SALT
1 TABLESPOON SEA SALT
2 CARDAMOM PODS, OPTIONAL

1. Preheat the oven to 300°F.
2. Scrub the beets thoroughly and prick each one in several places with a paring knife.
3. Place the beets in a baking dish and roast until they can be pierced easily with a knife, 2½ to 3 hours.
4. Once the beets are cool enough to handle, slip them out of their skins.
5. Cut the beets into fine julienne (the size of matchsticks).
6. Pack the beets in two 1-pint wide-mouth canning jars.
7. Mash the beets into the jars with a mallet or wooden spoon.
8. To make the brine, combine the water, whey, salt, and cardamom (if using) in a bowl and stir until the salt is dissolved.

9. Pour the brine over the beets, dividing it evenly between the two jars. Add more filtered water if necessary to completely cover the beets. The liquid should reach to about 1 inch below the top of the jars.
10. Cover the jars tightly and let the beets ferment at room temperature away from direct sunlight for about 3 days.
11. Store the pickled beets in covered containers in the refrigerator for up to 1 year.

Pickled Red Bell Peppers

MAKES TWO 1-PINT JARS

- WHEY STARTER (OPTIONAL)
- LACTO-FERMENT
- PREPARATION TIME FOR PEPPERS: ABOUT 1 HOUR
- FERMENT AT ROOM TEMPERATURE
- FERMENTATION TIME: 3 DAYS

Slow-roasting the peppers gives them a luscious texture, while fermenting adds a savory richness that fresh peppers never have. Use thick-fleshed peppers for this recipe.

12 RED BELL PEPPERS, QUARTERED AND SEEDED
½ CUP FILTERED WATER, PLUS MORE AS NEEDED
¼ CUP WHEY OR ADDITIONAL 1 TEASPOON SEA SALT
1 TABLESPOON SEA SALT

1. Preheat the oven to 325°F.
2. Place the peppers in a single layer on an oiled baking sheet.
3. Roast the peppers until the skin blisters and the peppers are very soft, 20 to 30 minutes. Turn the peppers to roast evenly on all sides.
4. Place the peppers in a bowl and cover with plastic wrap.
5. When the peppers are cool enough to handle, remove and discard the skins.
6. Pack the peppers in two 1-pint wide-mouth canning jars.
7. To make the brine, combine the water, whey, and salt in a large bowl and stir until the salt is dissolved.
8. Pour the brine over the peppers. Add more filtered water if necessary to completely cover the peppers. The liquid should reach to about 1 inch below the top of the jars.

9. Cover the jars tightly and let the peppers ferment at room temperature in an area away from direct sunlight for about 3 days.
10. Store the pickled peppers in covered containers in the refrigerator for up to 1 year.

Pickled Eggplant

MAKES TWO 1-PINT JARS

- LACTO-FERMENT
- FERMENT AT ROOM TEMPERATURE
- FERMENTATION TIME: 1 DAY

Enjoy this as part of an antipasti platter, or serve it on grilled bread. You can also serve it as a topping on sautéed, grilled, or roasted poultry, fish, or seafood. When you use pickled eggplant, lift it out of the jar with a fork or slotted spoon so the extra oil drains back into the jar. When you've eaten all the eggplant, use the flavored oil to drizzle on pasta or cooked vegetables, add it to salad dressings, or use it to sauté meats, fish, or vegetables.

4 MEDIUM EGGPLANTS, PEELED, SLICED, AND CUT INTO ¾-INCH-WIDE STRIPS
2 MEDIUM RED BELL PEPPERS, STEMMED, SEEDED, AND CUT INTO ¼-INCH-WIDE STRIPS
3 HOT FRYING PEPPERS OR CHILIES, STEMMED, SEEDED, AND CUT INTO ¼-INCH-WIDE STRIPS
1 CUP FILTERED WATER
1 TABLESPOON SALT
¼ CUP RED WINE VINEGAR
2 TABLESPOONS HONEY
12 GARLIC CLOVES, MINCED
1 TABLESPOON CHOPPED FRESH OREGANO
1 TABLESPOON CHOPPED FRESH THYME
1 CUP OLIVE OIL
2 TEASPOONS SALT-PACKED CAPERS
10 GREEN OLIVES, PITTED AND CHOPPED

1. Place the eggplant and bell and frying peppers in a pickling bucket or crock.
2. To make the brine, combine the water and salt in a large bowl and stir until the salt is dissolved.

3. Pour the brine over the vegetables.
4. Knead the eggplant with your hands until it releases some of its juices, about 10 minutes.
5. Cover with cheesecloth or a clean towel. Use string or rubber bands to hold it in place.
6. Put a plate on top of the eggplant and add enough weight so that it is completely submerged. (See page 20 for suggestions about how to weight down vegetables as they ferment.)
7. Let the vegetables ferment at room temperature in an area away from direct sunlight for 1 day.
8. Drain the vegetables and spin dry them in a salad spinner.
9. To make the seasoning mixture, combine the vinegar, honey, garlic, capers, olives, oregano, and thyme.
10. Add the seasoning mixture to the drained vegetables and stir until the pieces are evenly coated and moistened.
11. Pack the vegetables in two 1-pint wide-mouth canning jars.
12. Add enough oil to completely cover the vegetables.
13. Store the pickled eggplant in covered containers in the refrigerator for up to 1 year.

Pickled Ginger

MAKES TWO 1-PINT JARS

- WHEY STARTER (OPTIONAL)
- LACTO-FERMENT
- FERMENT AT ROOM TEMPERATURE
- FERMENTATION TIME: 3 DAYS

To peel ginger, use an ordinary teaspoon and scrape the edge of the spoon against the ginger. The skin will come off easily.

3 POUNDS FRESH GINGER, PEELED AND SLICED VERY THIN
1 CUP FILTERED WATER, PLUS MORE AS NEEDED
¼ CUP WHEY OR ADDITIONAL 1 TEASPOON SEA SALT
1 TABLESPOON SEA SALT

1. Pound the ginger with a meat pounder or mallet until it releases some of its juices, about 10 minutes.
2. Pack the ginger in two 1-pint wide-mouth canning jars.
3. To make the brine, combine the water, whey, and salt in a large bowl and stir until the salt is dissolved.
4. Pour the brine over the ginger, dividing it evenly between the jars. Add more filtered water if necessary to cover the ginger completely. The liquid should reach to about 1 inch below the top of the jars.
5. Cover the jars tightly and let the ginger ferment at room temperature in an area away from direct sunlight for about 3 days.
6. Store the pickled ginger in covered containers in the refrigerator for up to 1 year.

Corn Relish

MAKES TWO 1-PINT JARS

- WHEY STARTER (OPTIONAL)
- LACTO-FERMENT
- FERMENT AT ROOM TEMPERATURE
- FERMENTATION TIME: 3 DAYS

This relish is worlds better than any bottled relish you can find in the grocery store. The next time you have corn on the cob (ideally, in the summer at the height of the season when local corn is at its best), cook 3 or 4 extra ears. That will give you more than enough kernels for this relish.

3 CUPS CORN KERNELS
1 SMALL TOMATO, PEELED, SEEDED, AND DICED
1 SMALL ONION, MINCED
½ MEDIUM RED BELL PEPPER
½ CUP CHOPPED CILANTRO LEAVES
¼ TEASPOON RED PEPPER FLAKES
¼ CUP WHEY OR ADDITIONAL 1 TEASPOON SEA SALT
1 TABLESPOON SEA SALT

1. Combine the corn, tomato, onion, red bell pepper, cilantro, red pepper flakes, whey, and salt in a large bowl.
2. Pound the vegetables with a meat pounder or mallet until they release some of their juices, about 10 minutes.
3. Pack the corn relish in two 1-pint wide-mouth canning jars.
4. Press down firmly until the juices cover the top of the vegetables.
5. Cover the jars tightly and let the corn relish ferment at room temperature away from direct sunlight for about 3 days.
6. Store the corn relish in covered containers in the refrigerator for up to 1 year.

Garlic Dill Pickles

MAKES TWO 1-QUART JARS

- LACTO-FERMENT
- FERMENT AT ROOM TEMPERATURE
- FERMENTATION TIME: 3 DAYS

It is hard to believe that a process this simple can produce something as addictive as these crisp, tangy pickles. Kirbys are the best choice; they are the little, warty cucumbers you can find in great abundance from the middle to the end of the summer. In addition to being the right size, kirbys aren't waxed.

6 PICKLING CUCUMBERS
1 TABLESPOON MUSTARD SEEDS
4 MEDIUM GARLIC CLOVES
2 SPRIGS FRESH DILL
2 CUPS FILTERED WATER, PLUS MORE AS NEEDED
4½ TEASPOONS SEA SALT

1. Scrub the cucumbers.
2. Place the cucumbers in two 1-quart wide-mouth jars.
3. Divide the mustard seeds, garlic, and dill evenly between the two jars.
4. Combine the water and salt and divide evenly between the two jars.
5. Add more filtered water if necessary to completely cover the cucumbers.
6. Cover the jars tightly and let ferment at room temperature in an area away from direct sunlight for about 3 days.
7. Store the dill pickles in covered containers in the refrigerator for up to 1 year.

Tomato Salsa

MAKES TWO 1-PINT JARS

- WHEY STARTER (OPTIONAL)
- LACTO-FERMENT
- FERMENT AT ROOM TEMPERATURE
- FERMENTATION TIME: 2 DAYS

Letting the salsa ferment before you eat it adds a subtle richness to this popular relish. In addition to its popularity as a dip for tortilla chips, it is also a good topping for fish or chicken, chili (con carne or not), enchiladas, and burritos.

4 MEDIUM TOMATOES, PEELED, SEEDED, AND DICED
2 SMALL ONIONS, CHOPPED FINE
¼ CUP CHOPPED HOT CHILIES
6 CLOVES GARLIC
1 BUNCH CILANTRO, CHOPPED
1 TEASPOON DRIED OREGANO
JUICE OF 2 MEDIUM LIMES
¼ CUP FILTERED WATER, PLUS MORE AS NEEDED
¼ CUP WHEY OR ADDITIONAL 1 TEASPOON SEA SALT
1 TABLESPOON SEA SALT

1. Combine the tomatoes, onions, chilies, garlic, cilantro, oregano, lime juice, water, whey, and salt in a large bowl.
2. Transfer the salsa to two 1-pint canning jars.
3. Press the salsa to remove air pockets and to bring the liquid level above the top of the vegetables. Add more filtered water if necessary to completely cover the salsa. The liquid should reach to about 1 inch below the top of the jars.
4. Cover the jars tightly and let ferment at room temperature in an area away from direct sunlight for about 2 days.
5. Store the tomato salsa in covered containers in the refrigerator for up to 1 year.

CHAPTER 6

Fruit

Making fermented fruit chutneys and other relishes is no more difficult than making them from vegetables. Fruit ferments will not last as long as other types of ferments and they turn into alcohol. For some recipes, that is the goal. Adding sugar or alcohol into the dish adds flavor and also helps create an environment that discourages spoiling. In fact, a very famous fruit ferment has been popular for hundreds of years in Northern Europe. Fresh fruit is combined with sugar and some brandy. Equal parts sugar and fruit are used for the first batch, then the mixture is sweetened with honey and doused with dark rum or brandy. Each different fruit is added as it comes into season, and by Christmastime, there is plenty of delicious *rumtopf* for everyone to enjoy.

Pear Chutney

MAKES TWO 1-PINT JARS

- WHEY STARTER (OPTIONAL)
- LACTO-FERMENT
- FERMENT AT ROOM TEMPERATURE
- FERMENTATION TIME: 2 DAYS

½ CUP FILTERED WATER, PLUS MORE AS NEEDED
3 TABLESPOONS SUGAR
¼ CUP WHEY OR ADDITIONAL 1 TEASPOON SEA SALT
2 TEASPOONS SEA SALT
JUICE AND ZEST OF 2 MEDIUM LEMONS
3 CUPS DICED PEARS, PEELED, CORED, AND DICED
½ CUP TOASTED WALNUTS
½ CUP RAISINS
1 TEASPOON CUMIN SEEDS
1 TEASPOON FENNEL SEEDS
1 TEASPOON CORIANDER SEEDS
½ TEASPOON GREEN PEPPERCORNS
½ TEASPOON RED PEPPER FLAKES
½ TEASPOON DRIED THYME

1. To make the brine, combine the water, sugar, whey, salt, lemon juice, and lemon zest until the salt is dissolved. Add the pears to the brine. Toss to coat the fruit evenly.
2. Add the nuts, raisins, cumin, fennel, coriander, green peppercorns, red pepper flakes, and thyme. Pack the chutney into two 1-pint wide-mouth canning jars. Press down lightly with a mallet or wooden spoon. Add more filtered water if necessary to cover the fruit completely. The liquid should reach to about 1 inch below the top of the jars.
3. Let the chutney ferment at room temperature for 2 days.
4. Store the chutney in covered containers in the refrigerator for up to 2 months.

Spicy Pineapple Chutney

MAKES TWO 1-PINT JARS

- WHEY STARTER (OPTIONAL)
- LACTO-FERMENT
- FERMENT AT ROOM TEMPERATURE
- FERMENTATION TIME: 2 DAYS

This recipe provides ingredients for two fermented foods. The discarded skin and core can be used to make Pineapple Vinegar (page 97).

1 MEDIUM PINEAPPLE
1 BUNCH CILANTRO, CHOPPED
½ SMALL RED ONION, PEELED AND CHOPPED FINE
½ MEDIUM RED BELL PEPPER, SEEDED AND DICED
1 MEDIUM JALAPEÑO, SEEDED AND CHOPPED FINE
1 TABLESPOON COARSELY GRATED FRESH GINGER
½ CUP FILTERED WATER, PLUS MORE AS NEEDED
2 TABLESPOONS LIME JUICE
¼ CUP WHEY OR ADDITIONAL 1 TEASPOON SEA SALT
1 TEASPOON SEA SALT

1. Cut away the end and the outer skin of the pineapple. Cut the pineapple into quarters and cut out the core. Cut the pineapple into small pieces.
2. Combine the pineapple, cilantro, red onion, red bell pepper, jalapeño, and ginger in a large bowl and toss together.
3. To make the brine, mix the water, lime juice, whey, and salt until the salt dissolves.
4. Pack the chutney mixture in two 1-pint wide-mouth canning jars. Press down lightly with a mallet or wooden spoon. Divide the brine evenly between the jars. Add more filtered water if necessary to cover the chutney completely. The liquid should reach to about 1 inch below the top of the jars.
5. Let the chutney ferment at room temperature for 2 days.
6. Store the chutney in covered containers in the refrigerator for up to 4 months.

Preserved Lemons

MAKES 1 QUART

- LACTO-FERMENT
- FERMENT AT ROOM TEMPERATURE
- FERMENTATION TIME: 2 WEEKS

6 MEDIUM LEMONS, PREFERABLY ORGANIC AND THIN-SKINNED
¼ CUP SEA SALT
LEMON JUICE, AS NEEDED TO COVER THE LEMONS

1. To prepare the lemons, cut off the ends opposite the stem ends. Cut the lemons almost all the way through, into quarters, keeping them attached at the stem end. Sprinkle some of the salt between the sections.
2. Pour the remaining salt into a 1-quart wide-mouth canning jar. Push the lemons into the jar with a mallet or wooden spoon. Press down firmly until the juice is released. Add more lemon juice if necessary to cover the lemons completely. The liquid should reach to about 1 inch below the top of the jar.
3. Ferment the lemons at room temperature for 2 weeks, shaking and turning the jar daily.
4. Store the lemons in covered containers in the refrigerator for up to 4 months.

Bitter Orange Marmalade

MAKES TWO 1-PINT JARS

- WHEY STARTER (OPTIONAL)
- LACTO-FERMENT
- FERMENT AT ROOM TEMPERATURE
- FERMENTATION TIME: 3 DAYS

4 MEDIUM BITTER ORANGES
½ CUP FILTERED WATER
¼ CUP SUGAR
¼ CUP WHEY OR ADDITIONAL 1 TEASPOON SEA SALT
1 TABLESPOON SEA SALT
ORANGE JUICE, IF NEEDED

1. To prepare the oranges, quarter the oranges and slice them very thin. Remove and discard any seeds.
2. Divide the sliced oranges evenly between two 1-pint wide-mouth jars. Press down firmly with a mallet or wooden spoon until the juice is released.
3. To make the brine, combine the water, sugar, whey, and salt. Pour the brine over the oranges, dividing it evenly between the jars. Add orange juice or water if necessary to cover the oranges completely. The liquid should reach to about 1 inch below the top of the jars.
4. Let the marmalade ferment at room temperature for 3 days.
5. Store the marmalade in covered containers in the refrigerator for up to 2 months.

Note: If the marmalade develops white spots on the surface, lift them off with a spoon and discard them. The marmalade is still safe to eat.

CHAPTER 7

Dairy

Soured milk of all sorts has been around as long as we've had domesticated cattle. Before the means were invented to store fresh milk for long periods, milk would sour spontaneously after being left sitting out. Most of us can go to the store for fresh milk if what's in the carton pours out lumpy or thick, but before refrigeration, people had to figure out how to extract nourishment from all their food.

The benefits of yogurt as part of a healthy diet have been recognized for decades. We've all heard about the Eurasian Georgians who live to be more than one hundred years old on a diet that includes large quantities of yogurt. The living cultures in the yogurt help populate the gut with healthy bacteria that promote digestive health. Anyone undertaking a lengthy course of antibiotics, which tends to kill off the flora in the gut, is generally encouraged to eat plenty of yogurt. Some cultured dairy drinks, like kefir and buttermilk, fell out of favor except among health food fans, but with the growing interest in traditional foods and cooking, they are coming into their own.

Often, individuals who are lactose intolerant can still consume fermented dairy foods, since the microbes in the yogurt eat up the lactose in the milk, turning it into lactic acid. Since the intolerance is to the lactose rather than to lactic acid, there are no adverse reactions.

ABOUT CHEESE

Clifton Fadiman, the American essayist and editor, once famously described cheese as "milk's leap toward immortality." When you think about it, cheese is simply milk that has been allowed to ferment. At all the points along the way from fluid milk to solid cheese, the basic steps of fermentation are followed. Microbes consume the various nutrients they find in the food, turning it into lactic acid. This lowers the overall pH of milk. Acids or enzymes are also introduced to stimulate the milk's proteins to coagulate into curds. Salt is generally added to the cheese (although at different points in the process for different cheeses) to help control the process of fermentation as well as to act as a preservative and flavoring agent. The curds are drained, shaped, and aged (with or without the addition of molds by injection or as a rub on the surface). Each cheese has its own characteristics, and volumes have been written about cheesemaking and doctorates awarded to cheesemakers.

Natural cheeses, as distinct from processed cheeses, are living foods, like wines, and will continue to change as they age. The process of ripening and aging traditional cheeses is one that requires very specific conditions, often difficult to reproduce anywhere except in a specific geographic location. All the great cheeses of the world mature in a specific type of cave, showing how a microbial community can indelibly mark the foods that come from a certain region.

ENJOYING FERMENTED DAIRY FOODS

Adding a dairy product that already has some established cultures, such as yogurt or buttermilk, to regular milk or cream transforms the milk into something new. The results depend on the type of milk you choose; you can use whole or reduced-fat milks, combinations of milk and cream, or just cream.

Each of the recipes in this chapter produces a delicious food that is perfect to enjoy on its own. That doesn't mean you can't embellish them a bit. Some options are as simple as adding sliced fruit or a sprinkling of nuts. Yogurt can be piled into parfaits with granola. Kefir can be puréed with fresh fruit to make a smoothie. Crème fraîche can be added to sauces, curries, and soups to make them rich and mellow.

Why are there so many baking recipes that call for buttermilk, such as cornbread, biscuits, pancakes, and waffles? Soured milks have plenty of lactic acid. That acid, in conjunction with other ingredients in a batter, specifically baking soda, produce a delightful rise and a pleasing texture (or crumb). In addition, soured milks add a pleasantly acid note to baked goods.

Buttermilk, yogurt, and kefir make excellent marinades for meats, poultry, and fish. The lactic acid helps tenderize the flesh at the same time that it adds flavor.

Crème Fraîche

MAKES 1 PINT

- BUTTERMILK STARTER, PURCHASED OR HOMEMADE (PAGE 52)
- FERMENT AT ROOM TEMPERATURE
- FERMENTATION TIME: 18 TO 20 HOURS

This is one of the simplest fermented dairy items to make. All you need is a clean jar with a lid, some cream, and a little buttermilk. Buttermilk provides the culture that thickens the cream and gives it a slight tang. If you make your own buttermilk (see page 52), be sure to use that, but even commercially produced buttermilk makes a great crème fraîche. Add this to soups and sauces, or dollop it on top of anything you might top with sour cream.

2 CUPS HEAVY CREAM (NOT ULTRA-PASTEURIZED IF POSSIBLE)
1 TABLESPOON BUTTERMILK

1. Stir the cream and buttermilk together in a clean 1-pint jar.
2. Cover the jar and let the crème fraîche ferment in a warm, dark place, at about 70°F, until thick, 18 to 20 hours. Crème fraîche should smell sweet and creamy and have a slight tang and a buttery flavor.
3. It can be used immediately or stored in a covered container in the refrigerator for up to 2 weeks.

Note: Crème fraîche is often stirred into soups and sauces for added richness and a pleasantly sour flavor. It can also add extra richness and depth of flavor to your favorite cheesecake recipe.

Cultured Buttermilk

MAKES 2 QUARTS

- BUTTERMILK STARTER, PURCHASED OR FROM PREVIOUS BATCH
- FERMENT AT WARM ROOM TEMPERATURE
- FERMENTATION TIME: 14 TO 16 HOURS

If you know someone who makes her own buttermilk, ask if you can have some to use as a starter for your first batch. After that, you simply have to save a little bit to use as your own starter.

2 QUARTS WHOLE MILK
¼ CUP CULTURED BUTTERMILK, FROM PREVIOUS BATCH OR COMMERCIALLY PRODUCED

1. Stir the milk and buttermilk together in a clean 2-quart jar.
2. Cover the jar and let the buttermilk ferment in a warm, dark place, at about 80°F, until thickened and slightly curdled in appearance, 14 to 16 hours.
3. It can be used immediately or stored in a covered container in the refrigerator for up to 2 weeks. Reserve ¼ cup of the buttermilk as a starter for the next batch.

Greek-Style Yogurt

MAKES 2 QUARTS

- PLAIN YOGURT STARTER, PURCHASED OR FROM PREVIOUS BATCH
- INCUBATE AT 85 TO 95°F
- FERMENTATION TIME: 20 TO 24 HOURS

It can take a bit of experimentation to find the perfect amount of starter to add to your milk for the texture and flavor you like best. This recipe produces a thick, creamy yogurt, but if you prefer a denser yogurt, let the yogurt drain in a cheesecloth-lined colander set over a bowl. Save the whey, the resulting liquid, and store it in a covered jar in the refrigerator. Use it in the brine for lacto-fermented pickles.

2 QUARTS WHOLE MILK
3 TABLESPOONS PLAIN YOGURT

1. In a stainless steel pan, slowly heat the milk over low heat to 180°F.
2. Remove the pan from the heat and let the milk cool to 110°F. Gradually stir in the yogurt. Pour the mixture into a 2-quart jar. Cover the jar and let ferment at 85 to 90°F until thickened and smooth, 20 to 24 hours.
3. The yogurt can be used immediately or stored in a covered container in the refrigerator for up to 2 weeks. Reserve 3 tablespoons of the yogurt to make the next batch.

Cultured Butter and Buttermilk

MAKES 1 CUP BUTTER

This recipe produces both butter and buttermilk. The buttermilk made here is already cultured because you begin with a cultured cream, the crème fraîche. Use it as a drink, in cooking or baking, or as the starter for your next batch of Cultured Buttermilk (page 52).

1 QUART CRÈME FRAÎCHE (PAGE 51)

1. Whip the crème fraîche in a blender or food processor until clumps of butter begin to form.
2. Pour the crème fraîche into a colander lined with cheesecloth.
3. Let the liquid (buttermilk) drain into the bowl.
4. Transfer the butter to a stainless steel bowl.
5. Work the butter with a wooden spoon or paddle to press out as much buttermilk as possible.
6. Rinse the butter in cold water and form into a ball.
7. The butter can be used immediately or stored in a covered container in the refrigerator for up to 2 weeks.

Note: You may wish to add up to ½ teaspoon salt, as well as spices, herbs, or even grated hard cheeses to the butter for a delicious spread. Try adding minced garlic or chopped fresh parsley and chives.

Kefir

MAKES 1 QUART

- DAIRY KEFIR GRAINS
- FERMENT AT WARM ROOM TEMPERATURE, 75 TO 80°F
- FERMENTATION TIME: 18 TO 24 HOURS

Depending on the temperature and the length of the ferment, kefir can get a bit fizzy. If you like flavored kefir, purée the fully fermented kefir with fresh or frozen fruits to make a smoothie.

1 TABLESPOON KEFIR GRAINS (SEE NOTE)
2 CUPS WHOLE MILK (SEE NOTE), PLUS MORE FOR RINSING AND STORING KEFIR GRAINS
¼ CUP HEAVY CREAM (SEE NOTE)

1. Drain and rinse the kefir grains in cold water or milk.
2. Combine the milk and cream in a clear 1-quart canning jar.
3. Add the kefir grains.
4. Cover the jar and let ferment at 75 to 80°F until thickened with a pleasantly tart taste, 18 to 24 hours.
5. Use a strainer or slotted spoon to gently lift the kefir grains out of the kefir. Rinse them in a little cool milk. Transfer to a clean jar and add enough milk to cover.
6. The kefir can be drunk immediately or stored in a covered container in the refrigerator for up to 2 weeks.

Note: You can use any dairy milk, including cow's, goat's, or sheep's milk, as well as any nondairy milk, such as coconut, almond, or soy. If you are using nondairy milk, you should still store kefir grains in milk. If you are using kefir powder instead of kefir grains, you should be able to culture at least two, and possibly three batches by saving about 1/4 cup of the kefir and adding it to more milk and cream. Using cream is optional, and if you do not use it, the kefir will be a little less creamy, but still thick and flavorful. You can replace the cream with additional milk, or omit it completely. You can use kefir to make Cream Cheese (page 56).

Cream Cheese and Whey

MAKES 1 CUP AND ABOUT 3 CUPS WHEY

- BUTTERMILK STARTER, PURCHASED OR HOMEMADE (PAGE 52)
- FERMENT AT WARM ROOM TEMPERATURE, 75 TO 80°F
- FERMENTATION TIME: 14 TO 16 HOURS

This recipe will produce about 3 cups of whey. Save it to use in lacto-ferments, starters, or as a cooking liquid.

1 QUART WHOLE MILK
3 TABLESPOONS BUTTERMILK

1. Combine the milk and buttermilk in a clean jar.
2. Cover the jar and let ferment in a warm, dark place at 75 to 80°F, until thickened and slightly curdled in appearance, 14 to 16 hours.
3. Line a colander with cheesecloth or a jelly bag. Pour the cream cheese into the colander and let the whey drain from the cream cheese into the bowl until it stops dripping, 6 to 8 hours.
4. Tie the corners of the cheesecloth around the handle of a wooden spoon. Suspend the cream cheese over the strainer and let it continue to drain until all the liquid has drained out.
5. Scrape the cream cheese into a container.
6. The cream cheese can be used immediately or stored in a covered container in the refrigerator for up to 2 weeks.

Cottage Cheese

MAKES 1 PINT

- LEMON JUICE TO CURDLE MILK
- INCUBATE AT 85 TO 90°F
- FERMENTATION TIME: 10 TO 12 HOURS

2 QUARTS WHOLE MILK
2 TABLESPOONS LEMON JUICE

1. In a stainless steel pan, slowly heat the milk over low heat to 200°F.
2. Remove the pan from the heat and let the milk cool to 115°F.
3. Gradually stir in the lemon juice.
4. Cover the pan and let ferment at 85 to 90°F until large curds have formed and a yellowish liquid has separated from them, 10 to 12 hours.
5. Pour the curds and whey into a lined colander set in a bowl. Let the whey continue to drain into the bowl until the curds are drained to your preference, about 4 hours.
6. Scrape the cottage cheese into a container.
7. Store in a covered container in the refrigerator for up to 2 weeks.

Clabbered Cream

MAKES 1 CUP

- BUTTERMILK STARTER, PURCHASED OR HOMEMADE (PAGE 52)
- INCUBATE AT 80 TO 85°F
- FERMENTATION TIME: 12 TO 14 HOURS

Using raw milk is the only way to make traditional clabbered cream. However, a good approximation can be made from pasteurized milk by adding a culture, such as buttermilk or plain yogurt.

1 QUART MILK
2 TEASPOONS BUTTERMILK

1. Combine the milk and buttermilk in a 1-quart canning jar.
2. Cover the container and let ferment at 80 to 85°F until a thick curd forms on the surface and a thin yellow liquid separates from the curd, 12 to 14 hours.
3. Lift the curd from the liquid and transfer it to a jar or dish.
4. The clabbered cream can be used immediately or stored in a covered container in the refrigerator for up to 2 weeks.

Note: Clabbered cream has a texture and flavor similar to sour cream. Use it wherever you would use sour cream. Sweeten it lightly to serve with fruit or as a topping for pies or puddings.

CHAPTER 8

Beans

Fermented beans have a distinctive, savory flavor. You can enjoy them in the form of Chinese fermented black beans, in a variety of bean pastes, and as tofu, tempeh, or miso.

Making your own miso at home is certainly possible, as long as you are able to provide a relatively stable environment with even constant temperatures, as close to 60 or 65°F as possible. Making miso and other fermented bean items can be odorous, however, so you might also want to consider how close to your living quarters you want to keep it. But there are other delicious fermented bean dishes that you can create without a commitment of six months to one year before you can enjoy the fruits of your labor.

Beans are a protein-rich food and an excellent source of complex carbohydrates, vitamins, and minerals. Unfortunately, they also contain substances known as ogliosaccharides that can interfere with digestion. Some people are prone to uncomfortable and embarrassing gastric symptoms if they eat beans. When you ferment beans, either before they are cooked or afterward, you essentially predigest the components that cause gastric distress in our own digestive systems.

Beans that are fermented before they are cooked still have to be cooked, which means that the living organisms in the beans will be destroyed. The benefits of added flavor and easier digestibility are still there, however.

To ferment beans before cooking them, follow the steps for preparing any dry beans:

1. Sort through the beans, removing any dry, shriveled, or moldy ones.
2. Rinse the beans well in plenty of cold running water.
3. Put the beans in a large container (they will swell to nearly three times their dry volume as they soak).
4. Add enough cold water to cover the beans by 2 or 3 inches.
5. Cover loosely with a cloth or plastic wrap or with a lid and airlock.

6. Ferment at room temperature until you can see some bubbles forming, usually 24 to 36 hours.
7. Change the soaking water two or three times for the best flavor.

At this point, they are ready to be cooked in plenty of fresh water in your favorite recipe. You may find that they cook a little more quickly than beans that have not been fermented.

Fermented Tofu

MAKES 10 OUNCES

- LACTO-FERMENT
- FERMENT IN THE REFRIGERATOR
- FERMENTATION TIME: 9 TO 11 DAYS

The tofu is wrapped in a clean cloth before it ferments in a mixture of miso, mirin, and sake. You can use doubled cheesecloth or a thin cotton fabric known as voile. Be sure to rinse the cloth thoroughly in plenty of hot water before using it.

10 OUNCES MEDIUM OR FIRM TOFU, PREFERABLY ORGANIC OR NON-GMO
⅔ CUP WHITE MISO
4 TEASPOONS RED MISO
¼ CUP MIRIN
¼ CUP SAKE

1. Cut the tofu into 1-inch cubes. Drain on a paper towel–lined baking sheet for 15 to 20 minutes. Blot the tofu to remove any remaining surface moisture.
2. Combine the white and red miso, mirin, and sake. Spoon about one-third of the miso mixture into a container that will hold the tofu snugly.
3. Drape the cheesecloth or muslin over the miso mixture. Add the drained tofu in a single layer. Fold the cheesecloth over the tofu.
4. Smear the remaining miso mixture evenly over the cheesecloth to cover the tops and sides of the tofu.
5. Cover the container and let ferment in the refrigerator for 3 days. If liquid begins to collect in the container, drain it off and discard it.
6. Unwrap the tofu and transfer it to a clean container lined with parchment paper. Cover the container and transfer it to the refrigerator. Let the tofu age for 6 to 8 days.
7. The tofu can be used immediately or stored in a covered container in the refrigerator. Eat it within 2 weeks.

Tempeh

MAKES 10 OUNCES

- TEMPEH SPORE
- FERMENT AT WARM ROOM TEMPERATURE, 75 TO 85°F
- SOYBEAN PREPARATION TIME: 26 HOURS
- FERMENTATION TIME: 24 TO 36 HOURS

1 POUND DRIED SOYBEANS

2 TABLESPOONS RED OR WHITE WINE VINEGAR, CIDER VINEGAR, OR DISTILLED WHITE VINEGAR

1 TEASPOON TEMPEH SPORE (SEE RESOURCES, PAGE 127)

1. Soak the soybeans in a large container with enough cold water to cover them completely for 24 hours. Change the water two or three times for the best flavor.

2. Drain the soybeans and place them in a large pot. Add enough cold water to cover by 2 or 3 inches. Bring to a simmer over low heat. Simmer, skimming off any foam or hulls that float to the surface, until just barely tender to the bite, 1 to 1½ hours.

4. Drain the soybeans. When they are cool enough to handle, squeeze the skins gently and pop the soybeans out. Discard the skins.

5. Place the soybeans on sheet pans lined with paper towels and blot dry. Replace the damp towels with dry towels and let the soybeans air-dry at room temperature.

6. Combine the dried soybeans and tempeh spore in a large bowl. Transfer the soybeans to a 9 by 13-inch baking dish. Cut a piece of foil to cover the pan. Poke several holes in the foil and press it firmly onto the surface of the beans.

7. Let ferment in a warm place, 75 to 80°F, until the tempeh is covered with a white mold, 24 to 36 hours.

8. The tempeh can be used in recipes or cooled completely, cut into squares, and stored in a covered container in the refrigerator for up to 3 days. You can also steam the tempeh pieces (see note), cool them completely, and store in freezer bags or containers for up to 3 months.

Note: To prepare tempeh, cut the tempeh into pieces. Steam in a covered container over boiling water for 15 minutes. Tempeh can be grilled, broiled, baked, or sautéed, or added to your favorite soups or stews.

Miso

MAKES 1 PINT

- BARLEY OR RICE KOJI
- SOYBEAN PREPARATION TIME: 26 HOURS
- FERMENT AT ROOM TEMPERATURE
- FERMENTATION TIME: 6 TO 10 MONTHS

The barley koji provides the fungus you need to ferment soybeans into miso.

8 OUNCES WHOLE DRY SOYBEANS
1 POUND BARLEY KOJI (SEE RESOURCES, PAGE 127)
7 OUNCES SEA SALT (ABOUT 1 CUP)

1. Soak the soybeans in a large container in cold water to cover for 24 hours. Change the water two or three times for the best flavor.
2. Drain the soybeans. Place them in a large pot. Add cold water to cover by 2 or 3 inches.
3. Bring to a simmer over low heat. Simmer, skimming off any foam or hulls that float to the surface, until tender, 2 to 2½ hours. Ladle out 1⅔ cups of the cooking liquid and reserve.
4. Drain the beans. When they are cool enough to handle, squeeze the skins gently and pop the soybeans out. Discard the skins.
5. Place the beans on sheet pans lined with paper towels and blot dry. Replace the damp towels with dry towels and let the soybeans air-dry at room temperature.
6. Combine the dried soybeans and barley koji in a large bowl and mash until the soybeans turn into a coarse meal or paste.
7. With very clean hands, scoop the soybean mixture up and pack it into balls. Throw the balls into the fermenting container (this helps remove any air pockets). Press the soybean mixture firmly into the container.

8. Scatter the salt in an even layer over the surface of the soybeans. Press a piece of plastic wrap directly on the surface of the soybeans. Top with a 1-pound weight.
9. Ferment at room temperature, 65 to 70°F, until it has the flavor you like, 6 to 10 months. Shorter fermentations times produce milder, sweeter-tasting miso; longer fermentation times produce more intensely flavored miso.
10. As the miso ferments, a dark brown liquid will rise to the surface. This is tamari sauce. You can siphon or spoon it off (be sure your tools are scrupulously clean) to use as a condiment or in cooking.

Note: You can also use miso in any of the following ways:

- Stir the miso into simmering broth, adding it to your taste. Add cubes of silken or soft tofu and thinly sliced scallions and serve immediately.
- Add a spoonful to stews or braises just before serving to thicken the sauce slightly.
- Spread the miso on chicken, pork, beef, vegetables, or tofu as a rub before roasting or grilling.
- Add a spoonful to stir-fry dishes.
- Add a little to your favorite salad dressing, or blend miso with yogurt, rice wine vinegar, and a few drops of sesame oil to make a salad dressing or a sauce for vegetables or grains.

Dosa (Lentil Pancakes)

MAKES 6 DOSAS

- LACTO-FERMENT
- FERMENT AT ROOM TEMPERATURE
- LENTIL PREPARATION TIME: 24 HOURS
- FERMENTATION TIME: 2 TO 4 DAYS

1 CUP RED LENTILS
1 HOT GREEN CHILI, SEEDED AND CHOPPED
2 TABLESPOONS MINCED ONION
2 TABLESPOONS CHOPPED FRESH CILANTRO
1 TABLESPOON GRATED FRESH GINGER
1 TEASPOON KOSHER SALT
½ TEASPOON GROUND TURMERIC
¼ TEASPOON GROUND BLACK PEPPER
3 TABLESPOONS COCONUT OR OLIVE OIL

1. To prepare the lentils, combine them in a large bowl with warm water to cover. Soak at room temperature for 24 hours. Change the water two or three times for the best flavor. Reserve ½ cup of the soaking water.
2. To prepare the dosa batter, purée the drained lentils and the reserved ½ cup of soaking liquid with ½ cup water in a food processor or blender until smooth. Add the chili, onion, cilantro, ginger, salt, turmeric, and ground black pepper and continue to purée until smooth and creamy.
3. Transfer to a bowl. Cover with a towel and let ferment in a warm place, 75 to 85°F, until the batter smells tangy and is frothy, for at least 2 days and up to 4 days.
4. The batter can be used immediately or stored in a covered container in the refrigerator and used within 4 days.
5. To prepare dosas, preheat the oven to 180°F, to keep them warm as you make them.

6. Heat a griddle or skillet over medium-high heat. Add enough oil to generously coat the griddle or skillet. Add about 3 tablespoons of batter. Use the bottom of a spoon to spread the batter into a 6-inch-wide dosa. Make the dosas one a time, unless you have a large griddle that can hold two at once.
7. Cook the dosa until the edges start to look dry, 2 to 3 minutes. Drizzle about ½ teaspoon of the oil over the dosa. Turn the dosa and cook on the other side until done, 2 to 3 minutes.
8. Transfer to a sheet pan or baking dish and keep warm in the oven while making the remaining dosas. Add more oil to the griddle as necessary.
9. Serve the dosas warm.

Fermented Bean Dip

MAKES 1 PINT

- WHEY, HOMEMADE (PAGE 56) OR PURCHASED, OR KEFIR, HOMEMADE (PAGE 55) OR PURCHASED
- FERMENT AT ROOM TEMPERATURE
- FERMENTATION TIME: 2 TO 3 DAYS

2 CUPS COOKED BLACK OR PINTO BEANS
2 TABLESPOONS WHEY OR KEFIR
½ TEASPOON KOSHER SALT

1. Combine the drained beans, whey, and salt in a large bowl. Mash lightly to crush the bean skins a little.
2. Transfer to a 1-pint canning jar. Cover tightly and ferment at room temperature for 2 to 3 days.
3. The beans can be eaten immediately or stored in a covered container in the refrigerator for up to 2 weeks.

Note: Serve as a dip, in salads, or as an accompaniment to breakfast dishes like huevos rancheros, a warmed tortilla topped with beans, a fried egg, shredded cheese, salsa (page 43), and sour cream.

CHAPTER 9

Sourdough Bread

Sourdough bread has acquired the status of myth. Wherever sourdough bread is made—from the finest French bakeries to Western chuck wagons—sourdough starters are kept warm, coddled, nurtured, and fed regularly. There are stories that some sourdough starters (or "mothers") are hundreds of years old.

Whether you are a fan of San Francisco–style sourdough baguettes, dense peasant-style boules, or tangy sourdough rye breads, you already know that sourdough is the source not only of a satisfying, savory tang, but also of the bread's dense, moist crumb. It even makes bread last longer, without the preservatives found in the breads sold in supermakets.

You may have shied away from making your own sourdough starter because it seems so mysterious, not to mention sounding like a lifelong commitment and more demanding than many pets.

It is true that there is as much science as there is art to making sourdough bread. The science has to do with controlling the temperature, exposure to air, and amount of food available for the yeast. The art has to do with recognizing the sweet, beery aroma of a thriving starter and the color and feel of a starter in its prime, as well as the telltale hint of ammonia or the discolorations floating on the surface of a starter that has gone bad.

Starters that have turned color and won't come back to life when they are fed and left to ferment at room temperature cannot be saved. Sometimes the yeast runs out of food if you haven't been feeding it. Sometimes an unfriendly strain of yeast or other microbe can get established and take over; that's when you see colored streaks or get a whiff of a bad odor coming from the starter. The wonderful thing is, if at first you don't succeed, you can try again. The only thing you've really lost is a little flour and a few minutes of time.

You may have read that you need to use a little instant or fresh yeast to get a starter going, or some organic grapes or raisins. These tactics will introduce yeast, but you can also capture your own native wild yeasts. In time, the starter will develop a flavor that is as unique as your own home.

If you do make your own starter, you will probably end up with a lot of sourdough that you won't use. One obvious answer is to share it with your friends, but remember that some people won't be able to give it the attention it needs. Reassure your friends that, if they kill the starter, there's a good chance you'll have more for them very soon. And, if they turn down the offer, you can use your immature starter in the pancake or biscuit recipes in this chapter.

Rye or Wheat Sourdough Starter

- SOURDOUGH
- FERMENTATION TIME: 4 TO 6 DAYS

Look for organic whole-grain flours and use fresh, unchlorinated water to make your starter. You need to discard (or at least give away) about half the starter on each of the 5 or 6 days it takes to establish it. Not only does it keep you from ending up with enough starter to feed a small country, but it also makes a stronger, healthier starter, since there is less yeast competing for the fresh food.

Phase 1: Establishing the Starter

This phase takes approximately 5 days. If your kitchen is very warm, the starter might double by the second day, so it may be well enough established to use by day 4. If your kitchen is cool, however, you may need to add an extra day so that the wild yeast can really take hold. Maintain a regular feeding schedule while you are establishing the starter. If you are able to manage it, you may want to try feeding your starter twice a day during this stage. It's more work, but your reward is a more vigorous, robust starter.

DAY 1

1 CUP WHOLE-GRAIN RYE OR WHEAT FLOUR
½ CUP COLD FILTERED WATER

1. Combine the flour and water in a glass jar or an earthenware container.
2. Cover with a tightly woven cloth, secure with string or a rubber band, and leave at room temperature for 24 hours.

DAY 2

STARTER FROM DAY 1
1 CUP WHOLE-GRAIN RYE OR WHEAT FLOUR
½ CUP COLD FILTERED WATER

1. Stir the starter and put ½ cup of the starter into a jar or bowl.

continued ▶

2. Stir the flour and water into the starter. Cover with a tightly woven cloth, secure with string or a rubber band, and leave at room temperature for 24 hours. At this point, you may start to see some bubbles and the starter will increase in volume.

DAY 3

STARTER FROM DAY 2
1 CUP WHOLE-GRAIN RYE OR WHEAT FLOUR
½ CUP COLD FILTERED WATER

1. Stir the starter and put ½ cup of the starter into a jar or bowl.
2. Add the flour and water to the starter. Stir well, cover with a tightly woven cloth, secure with string or a rubber band, and leave at room temperature for 24 hours. The starter should increase in volume and may double.

DAY 4

STARTER FROM DAY 3
1 CUP WHOLE-GRAIN RYE OR WHEAT FLOUR
½ CUP COLD FILTERED WATER

1. Stir the starter and put ½ cup of the starter into a jar or bowl.
2. Add the flour and water to the starter. Stir well, cover with a tightly woven cloth, secure with string or a rubber band, and leave at room temperature for 24 hours. The starter should double in volume.

DAY 5

STARTER FROM DAY 4
1 CUP WHOLE-GRAIN RYE OR WHEAT FLOUR
½ CUP COLD FILTERED WATER

1. Stir the starter and put ½ cup of the starter into a jar or bowl.
2. Add the flour and water to the starter. Stir well, cover with a tightly woven cloth, secure with string or a rubber band, and leave at room temperature for 24 hours.

3. The starter should double in volume. Stir the starter down. It can be used for bread making immediately or stored in the refrigerator for 1 week.

Phase 2: Building the Starter

In this phase, you are feeding the starter with enough flour and water to compensate for the amount of starter you'll need for your recipe. If you have followed the instructions using the measurements given above, this "build" will give you enough to make two 1-pound loaves of bread.

THE BUILD FOR BREAD-BAKING DAY

2 CUPS WHOLE-GRAIN RYE OR WHEAT FLOUR
1 CUP COLD FILTERED WATER

1. If your starter has been in the refrigerator, stir it well to incorporate any liquid on the surface. Let the starter come to room temperature, about 3 hours.
2. Add 1 cup of the flour and ½ cup of the water to the starter. Mix well, cover, and let rise until doubled, 10 to 12 hours. Stir the starter down and measure out 1 cup of starter to use in your recipe (see pages 75–81).
3. Add the remaining 1 cup flour and ½ cup water to the starter.
4. Stir well, cover with a tightly woven cloth, secure with string or a rubber band, and leave at room temperature for 24 hours. The starter should double in volume.
5. The starter can be covered, refrigerated, and fed weekly at this point. If you plan to use your starter more frequently, you can keep it in a covered container on a counter at room temperature and feed it every other day.

Phase 3: Maintaining the Starter

You can keep the starter on the counter once it is established, if you are able to feed it once a day, or every other day, or if you plan to do a fair amount of baking over a period of a few days. Many people recommend keeping the starter in the refrigerator since they tend to bake less frequently. A refrigerated starter needs to be fed only once a week. Although you aren't likely to kill your starter if you aren't punctual to the hour, it is best to follow a fairly regular schedule.

continued ▶

Maintaining a starter is not the same as the building steps described in phase 2, but since you will feed only ½ cup of the existing starter, you'll have some starter to use for biscuits or pancakes (pages 81 and 79), to share with a friend, or, if all else fails, to put it in the compost bucket.

FEEDING DAY

STARTER
1 CUP WHOLE-GRAIN RYE OR WHEAT FLOUR
½ CUP COLD FILTERED WATER

1. Stir the starter and put ½ cup of the starter into a jar or bowl.
2. Add the flour and water. Stir well, cover with a tightly woven cloth, secure with string or a rubber band, and leave at room temperature for 24 hours. The starter should double in volume.
3. Stir the starter down. It is ready to build for use for bread making, or it can be stored, well-covered, in the refrigerator for 1 week, feeding it once a week. If kept on the counter, feed it every other day or daily.

Sourdough Rye Bread

MAKES 2 LOAVES

- SOURDOUGH STARTER
- FERMENT AT WARM ROOM TEMPERATURE, 75 TO 80°F
- FERMENTATION TIME: UP TO 3 HOURS BEFORE BAKING

If you do not have time to feed your starter before using it in this recipe, increase the yeast by an additional 1 teaspoon. The starter will still give the bread a delicious flavor and texture, but it will not be able to leaven the bread as effectively, which is why the recipe contains a little yeast. It gives the bread a good rise.

1 CUP RYE STARTER (PAGE 71)
2¼ CUPS WARM FILTERED WATER, 85 TO 90°F
3 TABLESPOONS HONEY OR MOLASSES
1 PACKAGE INSTANT YEAST
1 TABLESPOON KOSHER SALT
3 CUPS MEDIUM RYE FLOUR
3½ CUPS ALL-PURPOSE FLOUR
CORNMEAL FOR BAKING SHEETS
1 LARGE EGG WHITE
1 TABLESPOON CARAWAY SEEDS

1. Combine the starter and water in a large bowl, mixing with a wooden spoon until smooth.
2. Add the honey, yeast, and salt, and mix by hand until evenly combined.
3. Add the rye flour and 2 cups of the all-purpose flour. Continue to mix the dough by hand or in a stand mixer on low speed until the dough begins to pull away from the sides of the bowl, about 5 minutes by hand or 3 minutes in a stand mixer on low speed. It will still be quite wet at this point. Scrape the dough onto a floured work surface if mixing by hand.
4. Knead the dough, adding more of the remaining all-purpose flour as necessary to make a smooth dough, for about 7 minutes in a stand mixer on medium speed or for 10 minutes by hand.

continued ▶

5. Transfer the dough to an oiled bowl. Cover the dough with plastic wrap or a clean towel. Put the dough in a warm, draft-free place, 70 to 75°F. Let the dough rise until doubled, 1½ to 2 hours.

6. Fold the dough over on itself in three or four places to expel the gases. Divide the dough into two equal pieces. Shape the dough into round loaves.

7. Scatter a little cornmeal on two baking sheets. Place one round loaf on each baking sheet, seam-side down. Cover the loaves with plastic wrap or a clean towel, put in a warm place, and let rise until nearly doubled, about 30 to 45 minutes.

8. Preheat the oven to 400°F.

9. When the loaves have almost doubled in size, score the top of each loaf with a sharp paring knife, using diagonal slashes, into the top. The cuts should not be more than ¼ inch deep.

10. Brush the loaves with beaten egg white. Sprinkle the caraway seeds over the loaves, dividing them evenly between the two loaves.

11. Bake until the bread sounds hollow when tapped on the bottom or reaches an internal temperature of 190 to 195°F on an instant-read thermometer, 35 to 40 minutes.

12. Cool on a rack before slicing.

13. Wrap and store at room temperature.

Sourdough Baguettes

MAKES 3 BAGUETTES

- SOURDOUGH STARTER
- FERMENT AT WARM ROOM TEMPERATURE, 75 TO 80°F
- FERMENTATION TIME: UP TO 3 HOURS BEFORE BAKING

While loaves baked at home may not have the same crust and crumb as the ones from a bakery, you will still enjoy the tangy sourdough taste of these delicious baguettes. If you prefer a round loaf, divide the dough into two pieces instead of three, and shape into rounds.

1 CUP WHEAT SOURDOUGH STARTER (PAGE 71)
¾ CUPS WARM FILTERED WATER, 85 TO 90°F
2 TEASPOONS HONEY OR SUGAR
½ PACKAGE INSTANT DRY YEAST (OR 1¼ TEASPOONS)
2¼ TEASPOONS KOSHER SALT
5 CUPS ALL-PURPOSE OR BREAD FLOUR
CORNMEAL FOR BAKING SHEETS
1 LARGE EGG YOLK

1. Combine the starter and water in a large bowl, mixing with a wooden spoon until smooth. Add the honey or sugar, yeast, and salt and mix by hand until evenly combined.
2. Add 3 cups of the flour and continue to mix the dough by hand or in a stand mixer on low speed until the dough begins to pull away from the sides of the bowl, about 5 minutes by hand or 4 minutes in a stand mixer on low speed. It will still be quite wet at this point. Scrape the dough onto a floured work surface if mixing by hand.
3. Knead the dough, adding enough of the remaining flour to make a smooth dough, about 7 minutes in a stand mixer on medium speed or 10 minutes by hand.
4. Transfer the dough to an oiled bowl. Cover the dough with plastic wrap or a clean towel.

continued ▶

5. Put the dough in a warm, draft-free place, 70 to 75°F. Let the dough rise until doubled, 1½ to 2 hours.
6. Fold the dough over on itself in three or four places to expel the gases.
7. Divide the dough into three equal pieces. Shape the dough into baguettes as follows: Press one piece of dough into a rectangle, twice as long as it is wide and about ½ inch thick. Holding the sides of the dough, stretch it out until it is about 12 inches long and 6 inches wide. Starting at the top edge of the rectangle, roll the dough into a cylinder. Pinch the seam closed on the bottom and the roll the cylinder back and forth, pressing down and out as you roll, until the loaf is about 16 inches long. Tuck the ends under to the bottom of the baguette. Repeat with the remaining pieces of dough.
8. Scatter a little cornmeal on two baking sheets. Place the baguettes on the prepared baking sheets seam-side down. Cover the loaves with plastic wrap or a clean towel, put in a warm place, and let rise until nearly doubled, 45 to 50 minutes.
9. Preheat the oven to 450°F.
10. When the loaves have doubled in size, score the top of each loaf with a sharp paring knife, using diagonal slashes, into the top.
11. Brush the loaves with the beaten egg yolk.
12. Bake until the bread sounds hollow when tapped on the bottom or reaches an internal temperature of 190 to 195°F on an instant-read thermometer, 35 to 40 minutes
13. Cool on a rack before slicing.
14. Wrap and store at room temperature.

Sourdough Pancakes

MAKES 6 SERVINGS (18 PANCAKES)

■ SOURDOUGH STARTER

This recipe is the perfect way to use up any sourdough starter you might otherwise have to throw away. It may make more pancakes than you need but the extras can be stored in containers or freezer bags and kept in the freezer for up to 4 weeks. Reheat them in a toaster or microwave.

1¾ CUPS ALL-PURPOSE FLOUR
3 TABLESPOONS SUGAR
1 TEASPOON BAKING POWDER
½ TEASPOON BAKING SODA
½ TEASPOON KOSHER SALT
3 LARGE EGGS
2 CUPS WHEAT SOURDOUGH STARTER (PAGE 71)
1 CUP MILK
¼ CUP UNSALTED BUTTER, MELTED AND COOLED

1. Place the flour, sugar, baking powder, baking soda, and salt in a large mixing bowl. Whisk the ingredients together.
2. Beat the eggs in a small mixing bowl. Add the starter, milk, and butter, and mix by hand until well combined.
3. Make a well in the center of the dry ingredients. Add the wet ingredients to the dry and stir with a wooden spoon until evenly blended. Do not overmix, or the pancakes will be tough. It is okay if the the batter still has a few little lumps.
4. Preheat the oven to 180°F.
5. Heat a skillet or griddle over medium-high heat. Brush with melted butter or oil or spray with cooking spray.

continued ▶

6. Pour the batter onto the griddle, about 1/4 cup for each pancake. Leave about 2 inches between them so they can spread and can be turned over more easily.
7. Cook on the first side until you see small bubbles bursting in the center of the pancake, 2 to 3 minutes. Turn the pancake over and cook the other side until golden brown and cooked through, 2 to 3 minutes. Continue until all the batter has been used.
8. Hold the finished pancakes in the oven to keep them warm.
9. Serve the pancakes with toppings such as syrup, fruit butter, fresh fruit, whipped cream, or powdered sugar.

Sourdough Buttermilk Biscuits

MAKES 14 BISCUITS

■ SOURDOUGH STARTER

These biscuits are a great accompaniment to stews of all sorts. If you have any left over, the next day you can split them and toast them briefly to give them a fresh-baked flavor and texture.

2 CUPS ALL-PURPOSE FLOUR
½ TEASPOON BAKING POWDER
¼ TEASPOON BAKING SODA
½ TEASPOON KOSHER SALT
½ CUP CHILLED UNSALTED BUTTER, CUBED
1 CUP WHEAT SOURDOUGH STARTER (PAGE 71) AT ROOM TEMPERATURE
¾ CUP CULTURED BUTTERMILK AT ROOM TEMPERATURE (PAGE 52)

1. Preheat the oven to 425°F. Brush two baking sheets with butter or oil or spray with cooking spray.
2. Place the flour, baking powder, baking soda, and salt in a large mixing bowl. Whisk the ingredients together.
3. Add the butter to the flour and cut it in with a pastry blender or two table knives until it resembles coarse meal.
4. To prepare the wet ingredients, stir together the starter and buttermilk and add them to the dry ingredients. Mix until it forms a soft dough that comes away from the sides of the bowl.
6. Turn the dough out onto a floured work surface. Knead the dough 3 or 4 times, just until the dough holds together.
7. Using a floured rolling pin, roll out the dough to a ½-inch thickness. Using a 2-inch round cutter, cut out the biscuits, pressing the cutter straight down into the dough without twisting for the best rise and the flakiest biscuits.

continued ▶

8. Transfer the biscuits to the prepared baking sheets. Space the biscuits about 2 inches apart. Cover the biscuits with a clean towel or plastic wrap. Let the biscuits rise in a warm place, 70 to 75°F, until nearly doubled, about 30 minutes.
9. Preheat the oven to 425°F while the biscuits are rising.
10. Bake until golden brown on top, 10 to 12 minutes.
11. Serve the biscuits warm, accompanied by Cultured Butter (page 54), honey, fruit preserves, or jams.

CHAPTER 10

Meat, Fish, and Eggs

For some people, the idea of eating fermented meats, fish, and eggs is off-putting. They might be surprised to learn that many of their favorite meats are treated to some fermentation. Sometimes, the fermentation process is followed by a long period of air-drying to further preserve the meat. This is the procedure for such classic hams as prosciutto and Smithfield hams. In other instances, the meat is smoked or cooked once fermentation is complete, as with such favorites as bacon and pastrami.

Fermenting meats does call for scrupulous attention to cleanliness, as well as the ability to control both temperature and humidity for extended periods of time in the case of specialties such as salami or summer sausage. They get their tanginess from the lactic acid produced during fermentation. Most of us don't have the right cool, damp conditions in our cellars to manage the production of cured and dried sausages, but making corned beef or gravlax is well within the reach of home fermenting enthusiasts.

Fermented fish, whether in the form of classic Asian fish sauce or a traditional pickled herring, is another type of fermentation that calls for nothing more elaborate than fish, salt, cheesecloth or a jar, and spices.

Corned Beef

MAKES 3 POUNDS (6 TO 8 SERVINGS)

- LACTO-FERMENT
- FERMENT AT ROOM TEMPERATURE
- FERMENTATION TIME: 7 DAYS
- CURING TIME IN THE REFRIGERATOR: 2 WEEKS

This recipe does not include saltpeter, a curing agent that contains nitrates, which gives some cured meats a pink or reddish color even after they are cooked. Saltpeter is not essential, but if you prefer the look of bright pink corned beef, you can add about ¼ teaspoon saltpeter to your brine. But this is not a universal preferererence; classic Irish corned beef is grayish brown.

6 CUPS COLD FILTERED WATER
¾ CUP SEA SALT
½ CUP BROWN SUGAR
6 GARLIC CLOVES
6 BAY LEAVES
1 CINNAMON STICK
2 TABLESPOONS WHOLE CORIANDER SEEDS
2 TABLESPOONS WHOLE BLACK PEPPERCORNS
1 TABLESPOON JUNIPER BERRIES
4 CLOVES
3 POUNDS BEEF BRISKET

1. To make the brine, stir together the water, salt, and sugar until completely dissolved. Add the garlic, bay leaves, cinnamon, coriander, peppercorns, juniper berries, and cloves.
2. Trim the beef and place it in a fermenting container. Pour the brine over the beef. Add more cold water if necessary to cover the meat completely.
3. Add a weight to keep the meat submerged. Cover the container and add the airlock.
4. Let ferment at room temperature, 65 to 70°F, for 7 days.

5. Transfer the container to the refrigerator and ferment for 2 weeks.
6. To cook the corned beef, remove it from the brine. Place it in a deep pot and add enough fresh cold water to cover it completely.
7. Bring the corned beef slowly to a simmer over low heat. Simmer until thoroughly heated and tender, 2 to 2½ hours.
8. Slice the corned beef and serve.

Note: For a traditional corned beef dinner, add wedges of cabbage and peeled boiling potatoes to the pot during the final 40 minutes of simmering time. Serve with plenty of hot mustard and horseradish.

Pickled Eggs

MAKES 6 EGGS

- LACTO-FERMENT
- FERMENT AT COOL ROOM TEMPERATURE, 60 TO 65°F
- FERMENTATION TIME: 3 DAYS

2 GARLIC CLOVES, PEELED
3 SPRIGS FRESH DILL
6 HARD-COOKED EGGS, PEELED
1½ TEASPOONS KOSHER SALT
3 CUPS FILTERED WATER, PLUS MORE AS NEEDED

1. Place 1 of the garlic cloves and 1 sprig of the dill in the bottom of a 2-quart glass canning jar. Add the peeled eggs. Add the remaining garlic and dill.
2. Stir together the water and salt in a small bowl until the salt dissolves. Pour the salted water over the eggs. Add more filtered water if necessary to bring the liquid level within 1 inch of the top.
3. Seal the jars and let ferment in a cool, dark spot, 60 to 65°F, until you can see bubbles forming, about 3 days. Transfer the jar to the refrigerator.
4. The eggs can be eaten immediately or stored in the refrigerator for up to 2 weeks.

Beet Pickled Eggs

MAKES 6 EGGS

- LACTO-FERMENT
- FERMENT AT COOL ROOM TEMPERATURE, 60 TO 65°F
- FERMENTATION TIME: 3 DAYS

4 THIN SLICES YELLOW ONION
6 ALLSPICE BERRIES
6 HARD-COOKED EGGS, PEELED
3 CUPS BEET KVASS (PAGE 114)

1. Place 2 slices of the onion and 3 of the allspice berries in the bottom of a 2-quart glass canning jar.
2. Add the peeled eggs, the remaining 2 slices of onion, and 3 allspice berries.
3. Pour the kvass over the eggs. Add filtered water if necessary to bring the liquid level within 1 inch of the top.
4. Seal the jar and let ferment in a cool, dark spot, 60 to 65°F, until you can see bubbles forming, about 2 days. (If your kitchen is warmer than the suggested temperatures, you can let the eggs ferment completely in the refrigerator. It will take about 4 days.)
5. Transfer the jar to the refrigerator.
6. The eggs can be eaten immediately or stored in the refrigerator for up to 2 weeks.

Gravlax

MAKES 1 FILLET (TEN TO TWELVE 2-OUNCE SERVINGS)

- LACTO-FERMENT
- FERMENT AT ROOM TEMPERATURE
- FERMENTATION TIME: 6 HOURS
- FERMENTATION TIME IN THE REFRIGERATOR: 24 TO 30 HOURS

In traditional Scandinavian cooking, the liquid released by the salmon as it cures is used to make a mustard sauce.

1 CUP SUGAR
2 CUPS KOSHER SALT
2 BUNCHES FRESH DILL, CHOPPED
1 SALMON FILLET (ABOUT 3 POUNDS)
1¾ TABLESPOONS VODKA

1. To prepare the gravlax cure, combine the sugar, salt, and dill in a bowl, rubbing the dill into the salt and sugar to bruise it.
2. Remove the pin bones from the salmon. To check for pin bones, run your fingertip gently from one end of the fillet to the other along the seams in the flesh. Pull out any small bones you find with tweezers or needle-nose pliers.
3. Lay the salmon on a large piece of doubled cheesecloth. The piece should be large enough to wrap around the fish two or three times. Brush the vodka over the salmon.
4. Pack half the gravlax cure on the top of the salmon. Bring the cheesecloth over the salmon and turn the fillet over. Fold the cheesecloth back to expose the bottom side and pack it with the remaining cure.
5. Fold the cheesecloth around the salmon securely, folding in the sides to completely cover it. Wrap the salmon in plastic wrap completely.

6. Transfer the salmon to a shallow bowl or baking dish. Put a plate or a small cutting board on top of the salmon. Put a few cans or a brick on top of the plate to weight it down.
7. Let the salmon ferment at room temperature for 6 hours.
8. Finish fermenting the salmon in the refrigerator until it has a good flavor and a sliceable texture, 24 to 30 hours. Unwrap the salmon and scrape away the cure. Rinse to remove any remaining cure, and then blot dry with paper towels.
9. The gravlax can be carved immediately (see note).
10. Store the gravlax well wrapped in the refrigerator and eat within 2 weeks.

Note: To carve gravlax, use a long, thin knife. Make cuts at an angle and slice the salmon as thinly as you can. You can combine the scraps and ends with cream cheese to make salmon spread for toasted bagels or for an omelet filling.

Fermented Fish Sauce

MAKES 1 CUP

- LACTO-FERMENT
- FERMENT AT ROOM TEMPERATURE, 65 TO 75ºF
- FERMENT TIME: 3 TO 4 DAYS
- FERMENTATION TIME IN THE REFRIGERATOR: 4 WEEKS

This sauce has a history that goes as far back as eating. Records describe a form of it in ancient Greece, Rome (the famous garum), Constantinople, and classical China, Japan, India, and Southeast Asia. Today's distinctly tamer version is Worcestershire sauce.

1 POUND WHOLE FRESH SARDINES
4 TABLESPOONS SEA SALT
1 GARLIC CLOVE, PEELED AND SMASHED
2 FRESH BAY LEAVES
1 STRIP LEMON PEEL, 2 INCHES LONG AND ¼ INCH WIDE
2 TEASPOONS WHOLE BLACK PEPPERCORNS
2 TABLESPOONS WHEY OR PICKLE BRINE
1½ CUPS FILTERED WATER, OR AS NEEDED TO FILL THE JAR

1. To prepare the sardines, chop them into 1-inch pieces. Transfer to a clean 1-quart glass canning jar. Use a meat mallet, a potato masher, or a wooden spoon to mash the fish.
2. Add the salt, smashed garlic, bay leaves, lemon peel, and peppercorns. Add the whey.
3. Pour in enough water to cover the fish and fill the jar to bring the liquid level within 1 inch of the top. Cover the jar.
4. Let ferment at room temperature until there are plenty of bubbles in the jar, 3 to 4 days.
5. Transfer the jar to the refrigerator. Continue to ferment for 4 weeks.

6. Drain the fish sauce through a fine-mesh strainer into a clean container. Discard the solids.
7. The fish sauce can be used immediately or stored in clean bottles with tight caps. Use within 6 months.

Pickled Herring

MAKES 1½ POUNDS

- LACTO-FERMENT
- FERMENT AT ROOM TEMPERATURE
- FERMENTATION TIME: 24 HOURS

1½ POUNDS HERRING FILLETS, CUT INTO 1-INCH PIECES
1 MEDIUM YELLOW ONION, SLICED THIN
2 TEASPOONS KOSHER SALT
½ TEASPOON WHOLE BLACK PEPPERCORNS
½ TEASPOON YELLOW MUSTARD SEEDS
¼ TEASPOON WHOLE CORIANDER SEEDS
2 BAY LEAVES
2 TABLESPOONS WHEY (OPTIONAL)

1. Combine the herring, onion, salt, peppercorns, mustard seeds, coriander seeds, bay leaves, and whey (if using) in a large bowl, tossing until the ingredients are distributed evenly. Pack into a 2-quart glass canning jar.
2. Press the fish down with a mallet or wooden spoon. Pour in enough water to cover the fish and bring the liquid level within 1 inch of the top.
3. Cover the jar. Let ferment at room temperature for 24 hours.
4. The herring can be eaten now or stored in the covered jar in the refrigerator. Eat it within 2 months.

CHAPTER 11

Vinegar

Making your own vinegar might seem like a lot of work, especially since there are so many commercial vinegars to choose from. While there is no denying that some specialty vinegars, such as balsamic or sherry vinegar made by the solera method, would be hard to duplicate at home, making wine or fruit vinegars at home is simple once you have an established batch going. All it takes to establish your first batch is alcohol, a mother, a crock, a warm dark place, and patience.

Vinegar is produced when the ethyl alcohol in liquids like hard cider, perry, beer, or wine is converted to acetic acid. The bacteria responsible for this transformation form a soft shape that is called a "mother."

VINEGAR MOTHERS

The mother that you'll need for your own vinegars is really a slippery, slimy colony of microbes that can turn alcoholic liquids like wine, beer, or cider into vinegar. They do this by converting the sugars in the alcohol into acetic acid, the acid that gives vinegar its characteristic sharp bite.

You can buy mothers from beer-brewing and wine-making stores or online. Some people recommend using specific mothers for specific types of vinegar. Some mothers take longer to grow than others. That fact, plus the differences in temperature and other conditions from one place to another, make it difficult to put a precise time on how long your vinegar will take to brew. However, assuming an ambient room temperature that doesn't range more than 5 degrees above or below 70°F, you can expect to have a vinegar with a good flavor in 3 to 6 months from the time it starts to ferment.

If you have an active fermenting community or brewing enthusiasts in your area, someone may have a piece of mother to share. The mother should be in enough of the vinegar to keep it moist until you can add it to a batch. If you order online, the mother is shipped in vinegar with instructions for feeding and use of the mother before you incorporate it into your own batches.

Like sourdough starter, the mother will eventually take on unique characteristics because of the combination of yeasts and microbes peculiar to your environment. You can keep a vinegar mother alive almost indefinitely with proper feeding and care.

MAINTAINING AND SHARING A MOTHER

After you make your first batch of vinegar, the mother is dependably self-perpetuating. As long as it is fed and not exposed to sunlight or temperature extremes, each new batch of vinegar should produce a new mother. The older mothers won't hurt the flavor of batches that follow, although eventually, they will take up too much room in your jug or crock. In addition, they can block the spigot if you are using a crock with a spigot, so you may have to lift them out in order to draw off the vinegar.

To keep a batch of vinegar going almost indefinitely, use the following guidelines:

Once the mother is established, feed it periodically with 1 or 2 cups of wine. This will slow down the formation of vinegar but will keep the mother strong.

When your vinegar is done, take out no more than 50 percent of vinegar from your jar or crock; leave the rest behind to ferment the next batch.

Replace the volume you removed with an alcoholic liquid. Usually you will add the same type of liquid as you started with, but you can experiment by adding combinations (some white wine and some red, for example, or replacing part of the wine with fresh, unpasteurized fruit juices).

GROWING YOUR OWN VINEGAR MOTHER

You can grow a mother simply by pouring a bottle of red wine into a wide-mouth jar and letting it sit in a warm dark place, undisturbed, until you notice a gelatinous, gooey slime developing on the surface of the wine. That is the mother. Depending on the wine you use, the variety of microbes in the bacteria, and the temperature in your house, it can take anywhere from 2 weeks to 2 months to grow a mother. Since the bacteria need some oxygen to convert the alcohol into vinegar, cover the jar with cheesecloth to keep the fruit flies out while letting the air in.

As your first mother grows, you might see a layer of mold on the surface. When it first emerges, the mold looks like little white blossoms. The mold is not dangerous, but lift it off the mother with the edge of a spoon before it has a chance to establish itself and crowd out the microbes you want to keep in your mother. Once the mold gets the upper hand, your mother is finished.

To use a homemade mother for a batch of vinegar, lift the mother out of the liquid it's been sitting in and put it in a jar or crock. The wine you added to the jar will have turned into vinegar at this point, although it may not have the full flavor you want.

DRINKING VINEGARS

Not everyone gets excited about the idea of drinking vinegar, but this custom has a venerable history throughout Asia and England, as well as the countries colonized by England. Vinegars and their variations preserve valuable food resources, making them a natural for the fermenting repertoire. And they are delightful drinks that slake your thirst, with or without the addition of sparkling water or alcohol.

Vinegar has been credited with curing everything from diabetes to osteoporosis. Is it possible that something most people think of only in terms of salad dressing could be so good for you? Although there are few studies that show a specific relationship between vinegar and the various health benefits that have been claimed for it, there is plenty of anecdotal evidence from both contemporary and historical accounts. Some are being further investigated, especially those related to diabetes and the effect vinegar seems to have on insulin production.

Drinking vinegar, usually diluted with water, has been touted as a benefit for those suffering from osteoporosis. There is some evidence suggesting that drinking vinegar before eating foods that are high in calcium can help the gut absorb the calcium more effectively. That could be a benefit to those who suffer from osteoporosis, which is a softening of the bones that can result in fractures.

We do know that the acids in vinegars can change the environment in the gut. This has the effect of slowing down the rate at which carbohydrates are digested. That means that the amount of sugar released into the blood stream is slowed. Studies on how this may benefit diabetics are still being conducted.

If you do drink vinegar, always dilute it with water. Otherwise the acids in the vinegar can eat at your tooth enamel.

Although you may have heard that vinegar can provide a wide array of vitamins, minerals, and other beneficial nutrients, the USDA does not find that vinegar contains significant amounts of any of these ingredients. However, there are living microbes in raw, unfiltered vinegar. Even after you strain the vinegar through cheesecloth, there will still be bits of the mother in the vinegar floating around. That's why raw vinegars look cloudy and have bits of sediment swirling around on the bottom of the bottle. The floating bits are the source of the beneficial microbes that promote overall gut health. The sediment is evidence that there are microbes living and growing in the vinegar. Once the vinegar is pasteurized, they are killed.

SHRUB SYRUPS

The American colonists brought the English tradition of preserving fruits in vinegar and sugar with them. This sweet-tart vinegar was combined with water to make an effective thirst quencher. Shrubs were a prominent ingredient in the punches popular at the time. Today, they are being rediscovered and used in cocktails in trendsetting bars and restaurants.

Shrubs are made from equal amounts of fresh fruit, sugar, and vinegar. Berries are a popular base for shrubs, but tree fruits, herbs, and even vegetables can also be used.

PASTEURIZING VINEGARS AND SHRUBS

If you have enough room to store vinegars and shrubs in the refrigerator, there is no need to pasteurize them. But if you would like to bottle them and keep them in a cupboard or pantry, you may want to pasteurize them.

To pasteurize vinegars and shrubs:

1. Heat them gently to 180°F over low heat.
2. Pour the vinegar through a funnel into bottles.
3. Cap tightly.
4. Store in a pantry or cupboard away from direct sunlight.

Pineapple Vinegar

MAKES 1 QUART

- LACTO-FERMENT
- FERMENT AT ROOM TEMPERATURE
- FERMENTATION TIME: 2 OR 3 DAYS

The enzymes in fresh pineapple can produce a very active vinegar with plenty of effervescence. Open jars carefully to avoid explosions.

1 MEDIUM PINEAPPLE
½ TEASPOON RED PEPPER FLAKES
2 TEASPOON DRIED OREGANO OR 2 SPRIGS FRESH OREGANO
2 QUARTS FILTERED WATER

1. Cut the skin off the pineapple and transfer the skin to a 2-quart canning jar.
2. Cut the pineapple flesh lengthwise into quarters. Cut out the core and transfer the core to the canning jar. (Use the remaining edible pineapple in fruit salads, salsas, or other dishes.)
3. Add the red pepper flakes and oregano to the canning jar. Pour the water over the pineapple skin and core, adding enough to bring the liquid level within 1 inch of the top and to cover the pineapple completely.
4. Cover the jar and let ferment at room temperature (65 to 70°F) for 2 or 3 days.
5. Strain the pineapple vinegar into clear jars with screw tops or clamp lids.
6. Store in a covered container in the refrigerator for up to 2 weeks.

Red Wine Vinegar

MAKES 2 QUARTS

- VINEGAR MOTHER
- FERMENT AT WARM ROOM TEMPERATURE, 70 TO 75°F
- FERMENTATION TIME: 3 WEEKS TO 4 MONTHS

For best results, look for a sulfite-free, organically produced wine. Some vinegar makers say that wines that are on the fruity side make the nicest vinegars. Others say it really doesn't matter; they save tag ends of bottles to make vinegar or to feed the mother and keep it vigorous.

1 VINEGAR MOTHER AND VINEGAR (ABOUT 1 CUP)
2 QUARTS DRY RED WINE
2 CUPS FILTERED WATER

1. Place the vinegar mother and vinegar in a glass or ceramic crock or jar.
2. Carefully pour in the wine and water so that it runs down the sides of the jar or crock rather than pouring them directly on the mother.
3. Cover the jar or crock with cheesecloth. Secure with a string or rubber band. Place in a warm, dark area, 70 to 75 °F.
4. Let the vinegar ferment until it has developed a strong acidic aroma and a new mother has grown, at least 3 weeks and up to 4 months, depending on conditions in your kitchen and the acidity level you want.
5. If you have used a crock with a spigot, draw off about 1 quart of the vinegar. If you have used a jar, use a ladle to lift out the vinegar; take care not to disturb the mother.
6. Strain the vinegar through a cheesecloth-lined sieve. Pour the vinegar into clean bottles. Seal tightly.
7. Store at room temperature in a cupboard or pantry away from direct sunlight for 4 months or longer.

Note: Check the vinegar starting at 3 weeks by first smelling it, then tasting it. To smell the vinegar without burning your nostrils, position yourself so your nose is behind the jar, not directly over it. Use one hand to gently wave over the jar. When the vinegar is ready, it will smell very strong and extremely sharp. Continue to ferment the vinegar if the aroma is weak. If it smells like vinegar, taste it to see if you like the flavor. If it is not strong enough, let it ferment 1 or 2 weeks longer. You can store the finished bottled vinegar in the refrigerator if you prefer.

Variations:

WHITE WINE VINEGAR

Replace the red wine with a dry white wine.

CHAMPAGNE VINEGAR

Replace the red wine with champagne or other sparkling wine.

SHERRY VINEGAR

Replace the red wine with dry or sweet sherry.

CIDER VINEGAR

Replace the red wine with hard cider or perry.

MALT VINEGAR

Replace the red wine with beer.

Blackberry Shrub

MAKES TWO 1-PINT BOTTLES

- LACTO-FERMENT
- FERMENT AT ROOM TEMPERATURE
- FERMENTATION TIME: 5 TO 9 DAYS

2 CUPS BLACKBERRIES
2 CUPS SUGAR
2 CUPS RED WINE VINEGAR

1. Combine the blackberries and sugar in a large mixing bowl.
2. Crush them together using a wooden mallet, a potato masher, or clean hands.
3. Cover with cheesecloth or a clean towel, securing it in place with a string or rubber band.
4. Ferment at room temperature until the sugar is completely dissolved, at least 8 hours and up to 2 days.
5. Add the vinegar.
6. Cover with cheesecloth or a clean towel, securing it in place with a string or rubber band.
7. Ferment at room temperature until the shrub has a rich, sweet-tart taste, for at least 4 and up to 7 days.
8. Strain the shrub through a fine-mesh sieve or cheesecloth.
9. Transfer to clean jars or bottles and cap tightly.
10. Store in a covered container in the refrigerator for up to 2 months.

Variations:

CHERRY SHRUB

Replace the blackberries with pitted cherries. Use balsamic vinegar instead of red wine vinegar, if available.

STRAWBERRY RHUBARB SHRUB

Replace the blackberries with 1½ cups quartered strawberries and 1½ cups thinly sliced rhubarb.

CHAPTER 12

Alcoholic and Nonalcoholic Beverages

Brewing your own beverages is a rewarding and delicious enterprise. Some beverages, like fruit sodas, are ready in a matter of just a few days. Others, like mead or wine, demand a good bit more patience.

Some of the simple drinks in this chapter call for a ginger bug. This is a type of starter that you build, much as you do a sourdough starter. Ginger, sugar, and water are combined and left to ferment. This mixture is fed daily over 4 or 5 consecutive days, until the ginger bug is bubbling and smells like sharp beer or yeast. Adding this to any fruit juice and letting it ferment changes the flavor of the juice and adds a touch of effervescence.

A word of caution is needed here: Sodas can become quite feisty as they ferment. If you have them in capped bottles as they ferment, you may want to "vent" them a few times in the early stages of fermentation so they don't build up too much pressure. Any time you open a bottle of home brew, make sure you are prepared for it to foam up vigorously.

Other drinks in this chapter are brewed with a specific type of yeast. These strains have been purified in order to produce reliable results. You can find brewing yeasts for a wide variety of beers and wines. There are strains for cider and mead as well. See Reources on page 127.

EQUIPMENT, INGREDIENTS, AND BASIC TECHNIQUES

Most of the equipment you need for brewing beverages is already in your kitchen. Soft beverages like sodas are especially simple. You just need a container (glass canning jars work well), some cheesecloth, and some clean bottles for storing it.

Hard beverages require a few additional pieces of equipment, but they are similar to what you already use for making any fermented food:

- Airlocks
- Siphoning equipment (tubes, clamps, and bottle fillers)
- Vat or container for crushing grapes
- Large pots for cooking beer wort
- Fermenting containers for primary and secondary fermentation
- Floating thermometers
- Hydrometer to measure sugar and alcohol levels
- Nylon mesh straining bags
- Funnel
- Bottle brush
- Bottles and bottle caps or corks
- Sanitizing solution or tablets (see page 22 for information about sterilizing equipment)

You can also find additional tools and additives at a brewing supply store or online. To learn more about making beer or wine at home, see Reources on page 127.

Ginger Bug for Soda

MAKES 1 PINT

- LACTO-FERMENT
- FERMENT AT ROOM TEMPERATURE
- FERMENTATION TIME: 5 DAYS

This bug is like a sourdough starter. You need to feed it daily for 5 consecutive days before it is established and ready to use. Then you feed it weekly to keep it alive and replenish it when you use it to make your own sodas.

DAY 1

3 TABLESPOONS CHOPPED FRESH GINGER
3 TABLESPOONS SUGAR
3 TABLESPOONS WATER

Stir together the ginger, sugar, and water in a 1-pint glass jar with a tight-fitting lid. Cover and let ferment at room temperature for 24 hours.

DAY 2 THROUGH DAY 5

On days 2, 3, 4, and 5, feed the bug by adding an additional 3 tablespoons chopped and peeled ginger, 3 tablespoons sugar, and 3 tablespoons water. By the end of day 5, you should have about 2 cups of ginger bug.

The ginger bug can be used immediately, or you can store it in the refrigerator. Feed the bug weekly if you are storing it. This will also replace any bug that you used to make soda.

Ginger Ale

MAKES 2 QUARTS

- GINGER BUG
- FERMENT AT ROOM TEMPERATURE
- FERMENTATION TIME: 2 OR 3 DAYS

As the ginger ale ferments, gases build up in the bottle. Remove the caps or tops of the jars carefully. Sometimes there can be a lot of action when the top first opens.

2 QUARTS FILTERED WATER
½ CUP SUGAR
½ CUP CHOPPED FRESH GINGER
½ CUP LEMON OR LIME JUICE
½ CUP GINGER BUG

1. To make the ginger wort, warm the water, sugar, ginger, and lemon juice in a stainless steel pan over medium heat until the sugar dissolves. Transfer the wort to a crock or large canning jars. Let the wort cool to room temperature.
2. Add the ginger bug to the wort. Cover with cheesecloth and let it ferment at room temperature until frothy and effervescent, 2 or 3 days.
3. Strain the ginger ale into glass bottles with screw or clamp-style caps.
4. The ginger ale can be drunk immediately or stored in tightly capped bottles in the refrigerator for up to 2 weeks.

Fruit Soda

MAKES 2 QUARTS

- GINGER BUG
- FERMENT AT ROOM TEMPERATURE
- FERMENTATION TIME: 3 DAYS

¼ CUP GINGER BUG (PAGE 104)
2 QUARTS FRESH OR BOTTLED FRUIT JUICE (SEE NOTE)

1. Add 2 tablespoons ginger bug each into 2 clean 1-quart jars. Pour the fruit juice over the bug.
2. Cover with cheesecloth and let ferment at room temperature until frothy and effervescent, 2 or 3 days.
3. Strain the soda into glass bottles with screw or clamp-style caps.
4. The soda can be drunk immediately or stored in tightly capped bottles in the refrigerator for up to 2 weeks.

Note: Any of the following fruit or vegetable juices can be used, either freshly made or store-bought (even pasteurized).

- Beet juice
- Blood orange juice
- Blueberry juice
- Celery juice
- Coconut water
- Cranberry juice
- Cucumber juice
- Grape juice
- Lemon juice (diluted with water and sweetened)
- Mango juice
- Orange juice
- Raspberry juice
- Tomato juice
- Watermelon juice

For even more flavor, add herbs such as the following:

- Basil
- Cilantro
- Mint
- Parsley
- Tarragon
- Thyme

SCOBY and Starter Tea for Kombucha

MAKES 1 SCOBY AND 3 CUPS STARTER TEA

■ BUILDING TIME: 2 TO 4 WEEKS

If you are using purchased kombucha to grow a SCOBY, many brewers suggest a plain kombucha. Look for brands that are organic and include the words "raw" or "living" on the label. You'll find them in the refrigerated section where you purchase natural foods. SCOBY stands for "symbiotic culture of bacteria and yeast." This recipe makes enough SCOBY and starter tea to brew 2 quarts of kombucha, as well as enough to use for a SCOBY hotel (next page) that keeps the SCOBY alive and thriving for use in new batches.

1 CUP WATER
1 TEA BAG BLACK TEA
3 TABLESPOONS SUGAR
2 CUPS PURCHASED PLAIN KOMBUCHA WITH LIVE CULTURES

1. Bring the water to a boil and combine it with the tea bag in a teapot.
2. Steep for 10 minutes (the tea should be quite strong). Remove the tea bag. Stir in the sugar.
3. Pour the sweetened tea and kombucha into a 2-quart glass canning jar. Cover the jar with cheesecloth and secure with a rubber band or string.
4. Let ferment at room temperature until a gelatinous-looking mass forms at the top of the jar, about 2 weeks and up to 4 weeks.
5. You may see bubbles and tendrils attached to the sides and bottom of the SCOBY. These are good.
6. When the SCOBY is about ½ inch thick, it can be used for kombucha immediately or stored in a SCOBY hotel (see next page).

Making a SCOBY Hotel

The liquid that the original SCOBY grew in is known as "starter tea." Since you need only 1 cup for a 2-quart batch, you can use the remainder of the starter tea to store the SCOBY between batches.

Store the SCOBY in a glass canning jar completely covered with a layer of the tea it grew in (about 2 cups), tightly covered in a cupboard for up to 30 days. Check the SCOBY during storage, and if the SCOBY is in danger of drying out, add more kombucha (your own brew or purchased) to keep it covered and nourished. You can use this method to create a backup SCOBY.

With each batch of kombucha you brew after the first batch, a new SCOBY will grow on top of the previous one. This "baby" SCOBY can be left attached to its mother, but once your SCOBY is about 2 inches thick, you should separate the babies. Just peel them apart with impeccably clean hands. Put the babies into jars as described above to save as backups or to give to friends.

Use a tightly woven cloth or several layers of cheesecloth to cover the SCOBY when it is in storage to keep bugs, dust, and dirt out of the jar and let in the air that the SCOBY needs.

Kombucha

MAKES 2 QUARTS

- SCOBY
- FERMENT AT ROOM TEMPERATURE
- FERMENTATION TIME: 2 TO 3 WEEKS OR LONGER

You can make kombucha with black tea, green tea, or a combination of teas. However, your SCOBY won't find the nourishment it needs in herbal teas. You can add herbal teas to your brew for flavor. For more instructions on adding flavorings to kombucha, see the note on the next page.

2 QUARTS WATER
4 TEA BAGS BLACK TEA OR 2 TABLESPOONS LOOSE TEA
½ CUP SUGAR
2 CUPS STARTER TEA (FROM GROWING SCOBY) OR PURCHASED RAW LIVING KOMBUCHA
1 SCOBY, HOMEGROWN (PAGE 108) OR PURCHASED

1. Bring the water to a rolling boil in a stainless steel pot. Add the tea and turn off the heat. Steep until the water is room temperature. Remove and discard the tea bags. Stir in the sugar.
2. Pour the sweetened tea into a 2-quart glass canning jar. Carefully pour in the starter tea and SCOBY. Cover the jar with cheesecloth.
3. Let ferment at room temperature until a gelatinous-looking mass forms at the top of the jar, 2 to 3 weeks.
4. Taste the kombucha periodically as it brews. It is ready when it has the flavor you like. The shorter the fermentation time, the sweeter it will be. The longer it brews, the sourer it will become.
5. Transfer the SCOBY to a SCOBY hotel (see page 109).
6. Ladle the kombucha into glass bottles with screw or clamp-style caps.

7. The kombucha can be drunk immediately or stored in the refrigerator in tightly capped bottles for up to 2 weeks.

Note: If you like a little more fizziness in your kombucha, let it ferment at room temperature for 1 or 2 days before transferring it to the refrigerator. If you want to add flavors to your kombucha, you can do that after you have finished fermenting it. Some options include the following:

- Fresh ginger
- Fresh herbs
- Fruit juices or purées
- Herbal teas
- Spices

Hard Cider or Perry

MAKES 2½ GALLONS

- YEAST
- FERMENT AT COOL ROOM TEMPERATURE, 60 TO 65°F
- INITIAL FERMENTATION TIME: 3 WEEKS
- SECONDARY FERMENTATION TIME: 4 WEEKS
- AGING TIME: 4 WEEKS

The difference between hard cider and perry is what fruit you use. Hard cider is made from apple cider, and perry is made from pear cider.

1 PACKAGE OR 2½ TEASPOONS CIDER, WINE, OR CHAMPAGNE YEAST
2½ GALLONS SWEET APPLE OR PEAR CIDER, PRESERVATIVE-FREE, PREFERABLY UNPASTEURIZED, PLUS 1 CUP FOR MAKING STARTER
2 POUNDS BROWN SUGAR

1. Make a starter by combining yeast and 1 cup of the cider in a small bowl. Cover and let ferment at room temperature until frothy, about 1 hour. Transfer the starter to the refrigerator. Remove from the refrigerator 2 or 3 hours before brewing to allow the starter to come to room temperature.
2. Pour the remaining cider into large clean pots and heat to about 180°F. Simmer for 45 minutes. Add the brown sugar to the cider as it simmers.
3. Pour the cider into a sterilized fermenting bucket and cool to room temperature. Gently stir the starter into the cider. Cover the fermenting bucket and add the airlock.
4. Let ferment at room temperature until the bubbling in the airlock ceases and the yeast has settled to the bottom of the fermenter, about 3 weeks from the start of fermentation.
5. Siphon the cider into a second sterilized fermenting bucket, leaving behind the bottom layer of sediment. (The cider can also be bottled at this point; it will still be cloudy.) Cover and ferment to clarify the cider, about 4 weeks.

6. For still cider, siphon the cider into sterilized bottles as described on page 125.

7. Store in the refrigerator for up to 1 year.

Variation:

SPARKING HARD CIDER

1. For sparkling hard cider, combine 1 cup water with an additional ¾ cup honey or brown sugar in a saucepan over medium heat and stir until the sweetener has dissolved.

2. Pour the mixture into a sterilized fermenting bucket or a bottling bucket.

3. Siphon the clarified cider on top of the sugar. Stir gently to distribute.

4. Siphon the cider into sterilized bottles. Let ferment at room temperature for 4 weeks.

5. Transfer to the refrigerator. Drink within 1 year.

Beet Kvass

MAKES 2 QUARTS

- LACTO-FERMENT
- FERMENT AT ROOM TEMPERATURE
- FERMENTATION TIME: 3 TO 4 DAYS

Kvass is an earthy tonic rich with nutrients that have been shown to help the body cleanse itself of toxins and aid digestion. Some studies are investigating how a brew like kvass might be beneficial to cancer patients. But there's more to kvass than its health benefits. Enjoy it for its refreshing flavor as a drink or as a tart addition to a variety of salad dressings, soups, stews, and sauces.

3 MEDIUM BEETS
1 QUART FILTERED WATER, PLUS MORE AS NEEDED
¼ CUP WHEY (PAGE 56)
1 TABLESPOON SEA SALT

1. Peel and coarsely chop the beets. Transfer the chopped beets to a 2-quart glass or ceramic container.
2. Combine the water, whey, and salt and pour over the beets. Add more water if needed to cover the beets completely. Cover with a tight-fitting lid.
3. Ferment at room temperature until bubbly with a pleasant sour aroma, 3 or 4 days. Strain the kvass.
4. The kvass can be served immediately.
5. Transfer the kvass to clean jars and seal tightly. Store in the refrigerator for up to 2 months.

Note: Save some of the kvass to use in place of the whey for your second batch of kvass.

Honey Mead

MAKES 1 GALLON

- YEAST
- FERMENT AT COOL ROOM TEMPERATURE, 60 TO 65°F
- INITIAL FERMENTATION TIME: 4 WEEKS
- SECONDARY FERMENTATION TIME: 4 WEEKS
- AGING TIME: 1 WEEK

If you would like a drier mead with a higher alcohol content, you can increase the quantity of honey to 6 pounds.

1 GALLON FILTERED WATER, PLUS 1 CUP FOR SOFTENING THE YEAST
4 POUNDS CLOVER HONEY
1 PACKAGE OR 2½ TEASPOONS MEAD, WINE, OR CHAMPAGNE YEAST

1. Heat 2 quarts of the water in a large pot to 90°F. Add the honey. Stir constantly over very low heat until well blended, taking care not to burn the honey, 4 to 5 minutes. Remove from the heat.
2. Stir in another 2 quarts water and continue to stir occasionally until the mixture cools to room temperature, 65 to 75°F.
3. While the honey mixture is cooling, stir the yeast together with the remaining 1 cup water. Set aside to soften for about 20 minutes.
4. Transfer the cooled honey mixture to a sterilized 1-gallon fermentation bucket. Stir in the softened yeast.
5. Cover with an airlock and ferment until fermentation has nearly stopped, 3 to 4 weeks. There will be one bubble every 60 seconds in the airlock.
6. Siphon the mead into a sterilized secondary fermentation container, leaving as much sediment behind as possible. Cover with an airlock and ferment until the mead clears, 3 to 4 weeks.
7. Siphon the mead into sterilized bottles. Cap the mead tightly.
8. Age in the refrigerator at least 1 week. Drink within 1 year.

Blackberry Mead

MAKES 1 GALLON

- YEAST
- FERMENT AT COOL ROOM TEMPERATURE, 60 TO 65°F
- INITIAL FERMENTATION TIME: 3 TO4 WEEKS
- SECONDARY FERMENTATION TIME: 3 TO 4 WEEKS
- AGING TIME: 1 WEEK

10 OUNCES FRESH OR FROZEN BLACKBERRIES
1 CLOVE
1-INCH PIECE VANILLA BEAN, SPLIT LENGTHWISE
1 GALLON FILTERED WATER, PLUS 1 CUP FOR SOFTENING THE YEAST
4 POUNDS CLOVER HONEY
1 PACKAGE RED STAR PREMIER CUVÉE YEAST

1. To prepare the blackberries, thaw if frozen and crush with your hands or a potato masher.
2. Make a sachet by tying the clove and vanilla bean in a piece of rinsed cheesecloth. Leave a long tail on the string you use so you can remove it more easily later on.
3. Heat 2 quarts of the water in a large pot to 90°F. Add the honey. Stir constantly over very low heat to until well blended, taking care not to burn the honey, 4 to 5 minutes. Remove from the heat.
4. Stir 2 quarts of the water into the honey mixture and continue to stir occasionally until it cools to room temperature, 65 to 75°F.
5. While the honey mixture is cooling, stir the yeast together with the remaining 1 cup water. Set aside to soften for about 20 minutes.
6. Transfer the cooled honey mixture to a sterilized 1-gallon fermentation bucket. Add the softened yeast, blackberries and their juices, and vanilla and clove sachet.

7. Cover with an airlock and ferment until fermentation has nearly stopped, 3 to 4 weeks. There will be one bubble every 60 seconds in the airlock.
8. Taste the mead during this time. Remove the vanilla and clove when you are satisfied with the flavor.
9. Siphon the mead into a sterilized secondary fermentation container, leaving as much sediment behind as possible. Cover with an airlock and ferment until the mead clears, 3 to 4 weeks.
10. Siphon the mead into sterilized bottles and cap tightly.
11. Let the mead age in the refrigerator for at least 1 week. Drink it within 1 year.

Root Beer

MAKES 2½ QUARTS

- GINGER BUG
- FERMENT AT ROOM TEMPERATURE
- FERMENTATION TIME: 2 DAYS

You can buy root beer extract from brewing supply stores or online, but making your own is simple if you have a good source for the basic ingredients. They can be found at many well-stocked health or natural foods stores. Another excellent source is Mountain Rose Herbs (see Resources, page 127).

3 QUARTS FILTERED WATER
¼ CUP DRIED SASSAFRAS ROOT BARK
¼ CUP WINTERGREEN LEAF
¼ CUP DRIED SARSAPARILLA ROOT
1 PIECE LICORICE ROOT, 5 INCHES LONG, BROKEN INTO PIECES
1 VANILLA BEAN, SPLIT
1 CINNAMON STICK
3 WHOLE STAR ANISE
1 TABLESPOON CHOPPED FRESH GINGER
1 TABLESPOON DRIED DANDELION ROOT
1 TABLESPOON DRIED BURDOCK ROOT
1 TABLESPOON CRUMBLED HOPS FLOWERS
1 TABLESPOON DRIED BIRCH BARK
1 TABLESPOON DRIED WILD CHERRY TREE BARK
1 TEASPOON DRIED JUNIPER BERRIES
2 STRIPS ORANGE ZEST, 2 TO 3 INCHES LONG, ¼ INCH WIDE
1½ CUPS UNREFINED CANE SUGAR
1 TABLESPOON MOLASSES
½ CUP STRAINED GINGER BUG (PAGE 104)

1. To make the extract, bring 2½ quarts of the water to a boil in a large pot over high heat.
2. Add the sassafras root bark, wintergreen leaf, sarsaparilla root, licorice root, vanilla bean, cinnamon stick, star anise, ginger, dandelion root,

burdock root, hops flowers, birch bark, wild cherry tree bark, and juniper berries.

3. Simmer uncovered until the flavors are extracted, 20 minutes. Add the orange zest, sugar, and molasses. Stir until the sugar is dissolved and spices, roots, and zest are fully steeped, 10 minutes.
4. Remove from the heat and strain through a cheesecloth-lined strainer into a 1-gallon glass canning jar or plastic bucket. Stir in the ginger bug and the remaining 2 cups water. Cool to 98°F.
5. Stir in the ginger bug and transfer the root beer to sterilized bottles. Cap tightly and ferment at room temperature for 2 days.
6. Taste a sample from one of the bottles. If you like the flavor, stop the fermentation by putting the bottles in the refrigerator. If not, let the root beer continue to ferment for 1 day at room temperature and taste it again.
7. Transfer the root beer to the refrigerator and drink within 3 months.

Basic Beer

MAKES 5 GALLONS (ABOUT FIFTY 12-OUNCE BOTTLES)

- YEAST
- FERMENT AT COOL ROOM TEMPERATURE, 60 TO 65°F
- INITIAL FERMENTATION TIME: 1 TO 2 WEEKS
- SECONDARY FERMENTATION TIME: 2 TO 4 WEEKS
- AGING TIME: 1 WEEK

Beer brewing has always been popular with home enthusiasts. The procedure is not difficult as long as you have a source for the ingredients and equipment you'll need. Finding supplies is easy through the Internet, but if there is a brewing supply store in your area, it's worth the trip. They'll have a wide array of grains, amendments, malts, yeasts, and tools. The exact ratio of hops, the type of hops (or the blend of hops), even the type of yeast can be a matter of serious debate.

This brewing method is referred to as warm fermentation or "top-cropping" because as the yeast ferments at warm temperatures, a frothy layer rises to the top of the beer.

Cleanliness always matters whenever you are fermenting. In the case of beer brewing, you must take the extra step of sterilizing all the equipment for brewing and storing beer. That includes things you might not think of, like thermometers and tubing.

Using tubing to siphon the wort or beer into fermenting or storage containers makes it easier to get the beer out and leave any sediment behind. If the sediment gets swirled around in the beer, you might end up with hazy beer with a peculiar flavor.

6 POUNDS UNHOPPED PALE MALT EXTRACT
2¼ OUNCES HOP PELLETS
1 PACKAGE OR 2½ TEASPOONS LIQUID YEAST
⅔ CUP CORN SUGAR OR OTHER PRIMING SUGAR

1. To make the wort, place the malt extract and hops in a large pot. Add enough water to cover completely. Bring to a boil and boil for 1 hour. Remove the wort from the heat and cool to room temperature.

2. Siphon or transfer the wort to the plastic bucket or carboy. Add the yeast. Seal the bucket or carboy with the lid and airlock.

3. Ferment at room temperature, 60 to 65°F, until fully fermented, 1 to 2 weeks. The hydrometer reading of beer that is ready to be bottled should be about 1.008 for dark beers and 1.010 to 1.015 for light beers. You can also judge readiness by tasting a sample; it should not be sweet tasting and there should be little or no bubbling action in the beer at this point.

4. Siphon the beer into a sanitized container. Add the corn sugar and stir very gently to combine. Siphon the beer into bottles and cap tightly.

5. Age the beer in the refrigerator for at least 2 and up to 4 weeks before drinking.

Note: Beers may continue to develop their best flavors over the course of several months. Some brewers condition their beers for up to 6 months, or even longer. Like all fermented foods and drinks, your nose will tell you right away if the beer should not be drunk. Take notes about the beer as it ages to learn how beers behave in your brewing environment. You can use the information to fine-tune the process each time you brew.

Green Malt (Malted Grain)

MAKES 15 POUNDS

- FERMENT AT ROOM TEMPERATURE
- FERMENTATION TIME: 3 TO 6 DAYS

This process requires plenty of time, plenty of space, and the ability to control temperatures during the process of sprouting the grain. You could break the quantities up if you can't process all the soaked grain at once.

10 POUNDS BARLEY, WHEAT, OATS, OR CORN (WHOLE BERRIES)

1. Wash the grain in clean water to remove the chaff, which will mostly float to the surface. Drain and place in a covered container with enough water to cover the grain by 2 inches.
2. Soak for 8 hours, drain, and let stand for 8 hours without water. After 8 hours, you may see whitish bulges on the ends of the grain. These are emerging roots.
3. Spread the soaked grain over paper towels in large baking pans and place inside black trash bags sealed airtight to hold in the moisture and to keep out dust.
4. Let the grain sprout at room temperature until the main shoot is the same length as the grain, 4 to 6 days for barley and 3 days for wheat, oats, or corn.
5. To dry the malted grain, see the recipe for Pale Malt (next page).

Pale Malt

- DRY IN OVEN
- DRYING TIME: UP TO 3 DAYS

To successfully determine when the malt has the appropriate percentage of water, you'll need to weigh the malted grain before drying and after drying to determine whether you've removed enough water. Malted grains weigh 50 percent more than dry grains. The goal is to get down to about 2 percent water by weight, which means that the total weight of the pale malt should be almost the same as the amount of dry grain you began with.

15 POUNDS GREEN MALT (PREVIOUS PAGE)

1. Spread the green malt on baking sheets; the layers should not be more than ½ inch thick. Place over a heat source (heating pad, seedling pad, etc.) or in an oven with the pilot light or lights turned on, at 100 to 125°F, until the moisture content of wheat, corn, or oats is reduced to 2 percent to 6 percent (the grain will weigh 10 pounds), 2 or 3 days.
2. Barley requires final drying in the oven at a temperature of 140 to 160°F until the malt contains 12 percent moisture (it should weigh 12 pounds 10 ounces). This can take anywhere from 24 hours to several days.
3. While the malt is drying, turn it every 30 minutes. To dry the malt slowly, begin drying it at 140°F and raise the temperature a few degrees each day (but never going above 160°F) over a period of 5 to 10 days while the malt dries. The finished grain malt should be crunchy and slightly sweet.

Basic Wine

MAKES 1 GALLON

- YEAST
- FERMENT AT COOL ROOM TEMPERATURE, 60 TO 65°F
- INITIAL FERMENTATION TIME: UP TO 8 WEEKS
- SECONDARY FERMENTATION TIME: UP TO 3 MONTHS
- AGING TIME: 6 MONTHS FOR WHITE WINE AND UP TO 1 YEAR FOR RED WINE

Wine making is a highly evolved set of skills. You can earn advanced degrees in enology and viniculture and still spend a lifetime perfecting your craft. That does not mean that wine making is beyond the skill of a home fermentation fan. The process is simple: Grapes are crushed to release their juice. The wild yeasts on the grape skins set to work, gobbling up the sugars, growing, and reproducing all the while. When they consume the sugars, they convert them into alcohol. Eventually, the food supply runs out and the amount of alcohol kills off the yeasts and stops fermentation.

When making wine, check that your equipment is scrupulously sterilized and then rinsed clean. Ask at a wine supply store about special detergents and bleaches. It is best to clean and rinse your equipment immediately before using. You can use Campden tablets to make a sanitizing solution. Crush 4 tablets for 1 quart water.

18 POUNDS WINE GRAPES
1/16 TEASPOON POWDERED POTASSIUM BISULFITE OR 1 CAMPDEN TABLET, CRUSHED
1/5 PACKAGE OR 1/2 TEASPOON WINE YEAST; 1 PACKAGE WILL FERMENT 1 TO 6 GALLONS OF WINE)
SUGAR AS NEEDED

1. Sort the grapes, discarding any shriveled or moldy ones. Remove the stems.

2. Crush the grapes to release the juice (called must) into the primary fermentation container. Your hands will work as well as anything.

3. Insert the hydrometer into the must. If it reads less than 1.010, consider adding sugar. If you're adding sugar, first dissolve granulated sugar in filtered water (adding sugar helps boost low alcohol levels). Stir the must thoroughly.
4. Add the powdered potassium bisulfite, crushed and dissolved in a little water or grape juice, and stir into the must.
5. After 24 hours, add the wine yeast by sprinkling it over the surface of the must. Stir and then cover the primary fermentation bucket with cloth; allow the must to ferment for 7 to 10 days.
6. Over the course of days, fermentation will cause a froth to develop on top (the cap) and sediment to fall to the bottom (the lees).
7. Siphon the must through a funnel into sanitized glass secondary fermentation containers, leaving the cap and the lees behind. Fill to the top to reduce the amount of air in contact with the wine. Fit the container with an airlock. Allow the juice to ferment for 6 to 8 weeks.
8. Siphon the wine into a clean glass secondary fermentation container. Continue to siphon the wine into a sterilized fermentation container every 3 weeks until the wine runs clear, about 3 months.
9. Siphon the wine into bottles, leaving space for the cork plus about ½ inch or so of extra room. Insert the corks.
10. Store the wine upright for 3 days, then store the bottles on their sides. The ideal storage temperature for wine is 55°F.
11. Let the wine age to complete its flavor development. Red wine is usually aged for at least 1 year. White wine can be drunk after 6 months.

Resources

Equipment, Cultures, Starters, and Mothers

Amazon: www.amazon.com

Cultures for Health: www.culturesforhealth.com

King Arthur Flour: www.kingarthurflour.com

Local Harvest: www.localharvest.org

Mountain Rose Herbs: www.mountainroseherbs.com

Nourished Kitchen: Reviving Traditional Foods (information, equipment, ingredients, and recipes): http://nourishedkitchen.com

Williams-Sonoma: www.williams-sonoma.com

Raw Milk

Real Milk: www.realmilk.com/real-milk-finder

Brewing Supplies and Equipment*

Brew Your Own Brew: http://brewyourownbrew.com

Midwest Supplies Homebrewing and Winemaking:
www.midwestsupplies.com/homebrewing-equipment.html

Winemakers Depot: www.winemakersdepot.com

Wine Making Superstore: www.winemakingsuperstore.com

* Also search local listings for brewing supplies and brewing supply stores in your area.

Glossary

Anaerobic: Environments without oxygen. In fermentation, an anaerobic environment is necessary for breaking down carbohydrates and turning them into sugar.

Brine: A saltwater solution for pickling or fermenting that acts on food by drawing out the water from its cells and killing any harmful bacteria that might spoil the food.

Buttermilk: A cultured milk, traditionally the by-product of churning butter.

Carboy: A large glass container used for primary fermentation.

Clabbered cream: A cultured cream made by allowing cream to rise to the surface of raw milk.

Incubator: Any object or equipment that will help keep a fermented food at the desired temperature throughout the fermentation process.

Kimchi: Korean-style fermented cabbage; generally spicy.

Kefir: A fermented beverage made with kefir grains.

Koji: A fermented starter used to produce miso grown on rice or barley. It is responsible for breaking down the carbohydrates and sugars in food products.

Kombucha: A fermented drink that originated in ancient China. It is made from a SCOBY (see next page), tea, and sugar. It has a slightly tangy taste.

Kvass: A fermented Russian brewed drink made from rye bread or beets. It has a flavor that's similar to root beer or cola.

Lactic acid: Stops the growth of harmful bacteria that might spoil food. It's produced by *Lactobacillus* in fermented foods.

Lactobacillus: A bacteria that helps produce lactic acid from carbohydrates. It is responsible for turning starches into sugars and acids and is essential for the fermentation process.

Phytic acid: These antinutrients are naturally occurring in some grains and can prevent healthy minerals from being absorbed by the body.

Primary fermentation: Initial stage in brewed beverages; the beverage is typically removed from sediment by transferring it to a secondary fermentation container or bottles.

Probiotics: Microorganisms that are healthful for the body and gut and that occur naturally in foods.

SCOBY: Symbiotic colony of bacteria and yeast. It is an essential culture needed for kombucha making.

Starter: Another name for any prefermented product. Starter cultures can be purchased commercially or made at home.

Wort: In home brewing, the name for the beverage or soda mix before the starter has been added and initiated fermentation.

References

Azam-Ali, M. M. (1998). *Fermented Fruits and Vegetables: A Global Perspective.* FAO Agricultural Services Bulletin No. 134.

Fallon, Sally. (1999). *Nourishing Traditions: The Cookbook That Challenges Politically Correct Nutrition and the Diet Dictocrats.* New Trends Publishing, Warsaw, Ind.

Katz, S. E. (2003). *The Art of Fermentation: An In-Depth Exploration of Essential Concepts and Processes from Around the World.* Chelsea Green Publishing, White River Junction, Vt.

Katz, S. E. (2003.). *Wild Fermentation: The Flavor, Nutrition, and Craft of Live-Culture Foods.* Chelsea Green Publishing, White River Junction, Vt.

Lewin, A. (2012). *Real Food Fermentation: Preserving Whole Fresh Food with Live Cultures in Your Home.* Quarry Books, Beverly, Mass.

Index

E

F

G

H

L

M

N

O

P

R

S

T

U

17056843R00083

Made in the USA
Middletown, DE
04 January 2015